AN EXPERIMENTAL

I

SURGICAL

AN ESSAY AWARDED THE CARTWRIGHT PRIZE
FOR 1897

BY

GEORGE W. CRILE, A.M., M.D., Ph.D.

PROFESSOR OF THE PRINCIPLES OF SURGERY AND APPLIED ANATOMY IN THE CLEVELAND COLLEGE OF
PHYSICIANS AND SURGEONS; FORMERLY PROFESSOR OF PHYSIOLOGY IN THE MEDICAL
DEPARTMENT OF THE UNIVERSITY OF WOOSTER; ATTENDING SURGEON TO
THE ST. ALEXIS AND CLEVELAND GENERAL HOSPITALS

PHILADELPHIA
J. B. LIPPINCOTT COMPANY

George Washington Crile

An Experimental Research into Surgical Shock

ISBN/EAN: 9783337898366

Printed in Europe, USA, Canada, Australia, Japan

Cover: Foto ©berggeist007 / pixelio.de

More available books at **www.hansebooks.com**

CONTENTS

INTRODUCTION

CONSIDERING the great importance of the subject, it is a matter for surprise that there has not as yet been presented, so far as I am aware, an account of any considerable experimental research into surgical shock. The present research had progressed but little before it became apparent that its scope was so comprehensive, and involved so many unsettled problems in physiology, that a clearly satisfactory termination could not be hoped for. The aim has been to obtain graphic data furnished in the performance of various surgical operations or by the infliction of different injuries.

For the sake of clearness certain hypotheses have been advanced in a part of the research, with the understanding that they are subject to future change. In the main I have aimed to obtain the results bearing on the subject, without always assuming to be able to explain the mechanism of their production. The magnitude of the subject and the great number of unsolved questions of physiology involved must serve as my apology for the incompleteness of the research. All the experiments were performed on unselected dogs, in regular order as they were supplied by the laboratory servant.

Complete records of each experiment, including the tracings, description of the animal, of the experiment, and of the autopsy, have been preserved. Only a sufficient number of tracings to serve as illustrations have been published.

In all, one hundred and forty-eight animals have been subjected to experiment. I have not presumed to even touch upon all the phases of the

7

question, nor do I profess to have exhausted the possibilities of those receiving the most careful attention.

I must embrace this opportunity for acknowledging my very great indebtedness to Professor Victor Horsley for many valuable suggestions, and for permitting me to perform the first sixteen experiments in his laboratory in the University College, London.

My best thanks are also due to my associate, Dr. W. E. Lower, who rendered most valuable assistance throughout the remainder of the research, which was carried out in the Physiological Laboratory of the Cleveland College of Physicians and Surgeons.

AN EXPERIMENTAL RESEARCH

into

SURGICAL SHOCK

HISTORICAL ACCOUNT

AS TO THEORIES CONCERNING SHOCK

JOHN HUNTER, in 1784, was probably the first to describe shock.

William Clowes, in 1568, Wieseman, in 1719, and Garengeot, in 1723, recognized shock, and attributed it to the presence of some foreign body in the wound or in the blood.

Guthrie, in his work "On Gunshot Wounds," speaks of waiting "until the alarm and shock have subsided," and details a number of cases.

James Little, in 1795, was the first writer to use the word shock in the sense it is now employed.

Travers, in 1827, described a certain form of shock as "prostration with excitement."

Sir Astley Cooper, in 1836, described a number of cases of shock, and believed that shock to the nervous system might cause death without reaction.

J. A. Delcasse, in 1834, stated that the effects of violence were transmitted chiefly through the osseous system, whereby the living molecules were separated from each other, especially in the brain, spinal cord, and liver.

Erichsen, in his treatise in 1864, considered shock in railroad and other accidents to be due to the "sharp vibration that is transmitted through everything,"—the immediate lesion being, probably, of a molecular character.

Verneuil, after a thorough discussion of the question, concludes that commotion (shock) is a series of phenomena, more or less sudden, following a traumatism of the anatomical elements, tissues, or organs, characterized by a temporary excitation or depression of the functions of the parts injured, and provoking changes anatomically comparable to those one observes normally in the stages of activity and of repose.

In 1870, Goltz, of Strasburg, made his classical experiments on the frog. When this animal was suspended with legs downward and smartly tapped on the mesentery with the handle of a scalpel, the heart was suddenly arrested. After a few minutes it resumed its beat, but was paler than before; less blood was thrown into the aorta. When, however, the frog was placed in the horizontal position, normal color returned to the heart, and the usual amount of blood was again thrown into the aorta. These phenomena were due to the vasomotor paralysis caused by mechanical violence. This experiment has been used as the basis of one of the most commonly received theories of shock.

In 1873, F. Lauder Brunton, in his monograph on the pathology and treatment of shock, accepted Goltz's theory.

Eisemann believed that the injury causes a contraction of the blood-vessels of the brain, whereby its normal functions are diminished or destroyed, and that fainting and the depression stage of shock are closely related. After contraction of the vessels there is hyperæmia.

Hofmeister, in 1885, called attention to the great number of deaths from shock in abdominal operations. He regarded malnutrition of the heart, fatty degeneration, general weakness, and loss of blood as the important factors in its production.

Furneaux Jordan, in his "Surgical Inquiries," states that every case of shock, whether through the intervention of the nervous system or not, acts upon the central organ of circulation and diminishes its force, and that without impaired cardiac action shock is impossible.

Gross describes shock as "a depression of the vital powers induced suddenly by external injury, and essentially dependent upon a loss of innervation." He mentions as the most severe and fatal cases those that supervene upon injury to the great nerve-centres, whose effects may be no less disastrous and rapid than the destruction of life by lightning. He refers also to sudden deaths from blows upon the epigastric region,—those inflicting injury upon the solar plexus. He points out that there is a great idiosyncrasy of the individual in relation to shock, and refers to fainting produced by the pricking of a needle and the introduction of a bougie as familiar illustrations of shock from trivial causes. The extraction of a tooth has been known to cause death from shock. As internal causes he refers to perforation of the bowel, passage of a gall-stone or renal calculus, extravasation of urine, and apoplectic seizures.

Agnew says, "Considered from a clinical stand-point, it is evident, first, that the determining cause of shock must reach that portion of the nervous system from which the heart and lungs receive their motor endowments, for the feeble action of these organs is one of the first observed phenomena of shock. By associating together feebleness of the heart and paralysis of the walls of the blood-vessels we can explain the accumulation of blood in the deep venous trunks."

Savory, in "Holmes's System," says, "Death from shock . . . is the result of a sudden and violent impression on some portion of the nervous system, acting at once through a nerve-centre upon the heart and destroying its action."

As Warren has said, "Though the literature of the subject is considerable since it received a place in surgery, yet few writers attempt to define the nature of shock. Its pathology is usually passed over briefly, and the term may be said to have been employed indiscriminately to describe all cases of sudden death following injury without hemorrhage. In America and in England the condition has been regarded as a general depression of the nervous system, without any well-defined idea as to what the nature of the change was."

Blum explains shock as a reflex irritation of the pneumogastric nerve, causing arrest of the heart's action.

Fisher enlarges upon the conclusions of Goltz, and believes that a great mass of blood accumulates in the abdominal vessels during shock, and while this is its principal cause, there is also a paralysis of the whole vasomotor system.

Schneider adopts the theory of vasomotor paralysis (based especially upon the observations of Folk and Sonnenburg on the cause of death after burns), which he ascribes to, first, an over-stimulation, hence contraction of the blood-vessels, then later a dilatation, causing an unequal distribution of the mass of blood. Under these circumstances the heart is unable to force the diminished amount of blood returned to it through the dilated blood-vessels. Its own muscles are insufficiently supplied with oxygen, and its action is greatly impaired. The symptoms of shock are all explained by the diminished blood-pressure.

Mansell Moullin is unable to admit that the views of Goltz, Fischer, and Schneider, although they must be regarded as an enormous advance upon all previous views, are a sufficient explanation of shock. In conclusion he says, "Shock is an example of reflex paralysis in the strictest and narrowest sense of the term, a reflex inhibition, probably in the majority of cases generally affecting all the functions of the nervous system and not limited to the heart and vessels only."

Groeningen, who has written probably the most complete treatise on this subject, offers his so-called neuro-hypothesis. Thus far the shock theories have had in the main a hemo-pathological basis, which affords a sufficient explanation of some phenomena. Groeningen then quotes Sir Astley Cooper as follows : " Extensive injuries cause death by their sympathetic influence and the great shock on the nervous system, even if there be no vessel reaction nor inflammation." This is, in a nut-shell, what can be said from a neuro-pathological stand-point. Shock is a lassitude or relaxation of the spinal marrow and of the medulla oblongata produced by violence. When a nerve is irritated its equilibrium is disturbed, and this disturbance is transmitted peripherally and centripetally. What becomes of the centrifugal disturbance we do not know. The centripetal produces a change in the central organ called sensibility or perception. In the nerve, as well as in the central organ, this action causes a consumption of matter, possibly also a change of position of the molecular elements. At any rate, after repeated irritations a chemical change in the nerves and a decrease of irritability is objectively established. The activity of the nerves is followed by a diminution of vitality called languor or exhaustion. The degree of exhaustion will be in proportion to the duration and the violence of the lesion. Four stages or degrees of irritation may be distinguished: *First.* The lowest is ineffectual. It does not pass beyond consciousness. *Second.* That which gives rise to the perception of the sense of touch, of sight, of hearing, of taste, or of smell. *Third.* A stronger irritation increases more or less the acuteness of these perceptions and brings out the sensation of pain or such sensations as loathing, etc. *Fourth.* The highest degree of irritation destroying all sensibility, either temporarily or permanently. These various degrees of irritation lead insensibly from the lower to the higher. A strong light may blind, an intensely loud noise may deafen. Of two or more simultaneous sensations, the stronger prevents the perception of the weaker. The lips are bitten to avoid the pain of an operation. A maximum irritation may destroy all the special senses, and even pain may not be felt.

The observation of Levisson on the results of pressing the kidney,— viz., paralysis of both hind paws and the abolishment of all reflex action in the hind limbs for some time after the pressure had ceased,—and other like observations, are explained by Goeningen as due to the effect of the violent irritation of the sensory nerves, temporarily causing paralysis of the motor apparatus in the spinal cord and exhaustion of the reflex centres, as well as a paralysis of the automatic centres in the brain.

A violent and sudden irritation of the peripheral or the sympathetic nerve fibres causes in the spinal cord an excitation, which is rarely confined to the territory of the affected nerve. Usually it is distributed over large sections of the spinal cord, and can therefore cause at the original point of lesion, as well as at more distant points, a condition of exhaustion and unconsciousness. This is produced by the momentary consumption of the nerve energy at hand.

In his excellent work on "Surgical Pathology and Therapeutics," Warren, after a critical review of the various theories of shock, advances still further its pathology by making use of the investigations of Hodge on the microscopical alterations in nerve-cells at rest and in fatigue. Hodge subjected the spinal ganglia of dogs and cats to electrical stimulation for some hours, then compared the changes observed in the cells with normal cells and the stimulated cells after a period of rest. His studies included the ordinary fatigue in swallows, bees, etc. He showed that metabolic changes are as easy to demonstrate as similar processes in gland-cells. Fatigue is characterized by a decrease in the size of the nucleus, and, instead of the smooth and rounded contour, an irregular outline obtains, the nucleus taking a darker stain; there is also a slight shrinkage in the size with vacuolation of the cells of the spinal ganglia, and considerable shrinkage of the cells of the cerebrum and the cerebellum with enlargement of the pericellular lymph-space. These observations seem to throw an additional light into the dark subject of the pathology of shock, and render it probable that in these profound functional disturbances similar changes may occur which may gradually disappear after repose.

CASES ATTENDED BY SHOCK

Moullin, under this head, refers to the great shock caused by burns and scalds, especially in cases involving extensive areas, by contused and lacerated wounds, and by capital operations, in all of which the danger is in proportion to the proximity of the injury to the trunk. Pirogoff saw two men die during the sawing of the femur in amputation of the thigh. A "spasmodic contraction passed over the muscles of the body, the face became pale, the eyes lost their lustre, the pupils dilated, and death followed at once."

Furneaux Jordan observed a drop of as much as one-fifth degree in the axillary temperature during the application of the saw. Injuries of the brain are by most surgeons considered causes of profound shock. Blows upon the scrotum or the pit of the stomach frequently cause great prostration or death. Fischer reports the case of a healthy man whose testicle was seized between the teeth of an enraged horse and severely lacerated, causing death in a few hours. Erichsen mentions the sinking of the pulse, even when the patient is fully under the influence of anæsthesia, at the moment of the division of the spermatic cord.

Sir Astley Cooper relates a case in which a laborer received a slight blow upon the epigastrium while wheeling a barrel and fell down dead. A number of instances are reported in surgical literature of sudden death following blows upon the epigastrium. Penetrating wounds of the abdomen, especially injuries and operations involving the handling of the viscera, are likely to be followed by shock. Strangulation of the small intestine and the result of the application of taxis in reducing hernia, may produce profound shock. Perforation of the stomach or of the intestine and parturition, especially in case of twins, are potent causes. Loss of blood especially predisposes to shock. According to Billroth, the evulsion of an arm or leg is usually followed by fatal shock.

Groeningen refers to injuries of the larynx as frequent causes of profound shock. Severe blows upon the neck—tactics often employed by prize-fighters—may be followed by great depression and unconsciousness. Fischer assumed that these injuries caused a spasm of the glottis, Bernard and others that an inhibitory action is exerted upon the respiratory centre through an irritation of the superior laryngeal nerve. Maschka reported the case of a boy twelve years of age who fell lifeless to the ground upon the receipt of a blow from a stone upon the larynx. Warren attributed the sudden death in this case in part to shock and in part to cerebral anæmia.

Age and sex seem to have an influence in producing shock, and most authors refer to the idiosyncrasy of the individual, the state of health, and the nervous organization as influencing the amount of shock following operations or injuries.

MODES OF INVESTIGATION AND ANNOTATION

In all cases the animals were anæsthetized, usually by the use of ether, occasionally by chloroform, either alone or with ether. In a few cases curare and morphine were used. The first administration was much

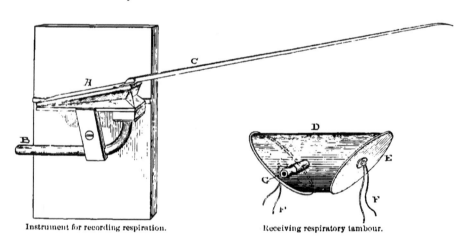

Instrument for recording respiration.   Receiving respiratory tambour.

facilitated, and the process relieved of much of its disagreeable detail, by thrusting cotton wool into the apex of a conical-shaped hood of canvas, with a sufficient capacity for receiving the dog's head and having a projection above for holding it in position, while the dog was held in a box allowing the forelegs to project over its top. On completion of anæsthesia tracheotomy was performed, and a simple anæsthetizing apparatus, consisting of a glass canula, to which was attached a rubber tubing, connected at its end with a funnel, was securely tied into the trachea. This arrangement not only facilitates the administration of the anæsthetic and secures free and unimpeded breathing, but also serves, by merely disconnecting the funnel and attaching the rubber tube to a connection with a bellows, as part of a simple and efficient artificial respiration apparatus. The latter apparatus consists of a bellows fastened to the under surface of the table, having a rubber tube with a glass canula for ready attachment to the breathing-tube above described. By means of adjusted weights for opening the bellows, and a rope over pulleys carried to the foot of the table, and thence to a hinge foot-board fastened to the floor, artificial respirations were easily maintained.

Respiratory movements were recorded by means of a broad canvas band encircling three-fourths the circumference of the dog's lower chest, to each end of which was clamped, under the necessary tension, the ends of strong threads (F) fastened to a small circular disk on the inside of the rubber

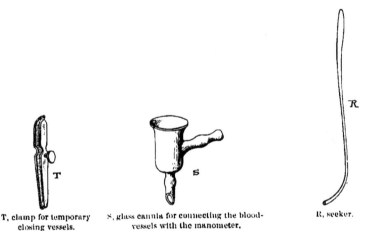

T, clamp for temporary closing vessels. S, glass canula for connecting the blood-vessels with the manometer. R, seeker.

dam, which was snugly drawn over the oblique end (E) of the three-inch air-tight brass tube. The latter, by this tension on each end, was held upon the sternum or upon the anterior aspect of the abdomen. The ends were cut oblique so as to make the respiratory actions operate more nearly at right angles to the plane of the elastic rubber. To the small tube (G) in the brass tambour was attached a rubber tubing, which acted upon a delicate recording instrument made on the principle, and of the material of an organ key. The rubber tubing was attached to a lead tube (*B*), communicating with the interior of the wedge-shaped "key" (*A*), which consists of a light wooden base and a like top, to which is cemented organ key leather, properly creased for expansion and contraction. On the upper surface of the wooden top is glued a small grooved piece of wood, and

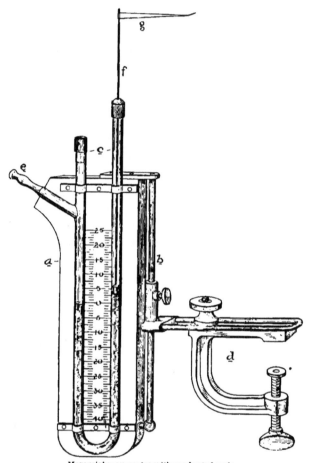

Mercurial manometer with graduated scale.

into the groove is pressed a long, slender, wooden, writing-style (*C*), armed at its end with a tapering, thin piece of Swedish spring steel. This apparatus is so delicate as to record even the heart impulses against the chest. The receiving apparatus was fastened to a light piece of board, which, in turn, was mounted on a burette-stand, making adjustment easy by simply pressing down or raising up this board in the grasp of the burette-clamp. The up-stroke represents expiration, the down-stroke inspiration.

CIRCULATION.—In the first sixteen experiments which were carried out in the University College, London, the records were obtained by means of a Hürthle kymograph carrying three metres of paper. Mer-

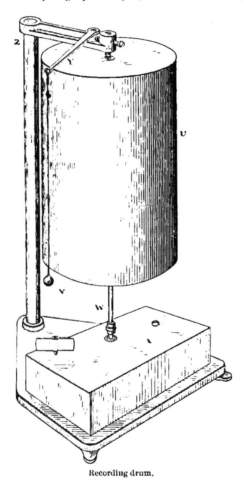

Recording drum.

curial manometers were used for the central blood-pressure, and for some of the peripheral pressures; in the greater number of the latter water-manometers were used. The writing-styles of the manometers and of the respiration apparatus were arranged in a vertical line at the beginning of the experiment. In using the water-manometers the record was traced by a float made of a series of truncated cones of cork held together by a

2

wire passing through their centre and projecting several inches below the lower cork. To the free end of this wire sealing-wax was attached as ballast. The edges of the corks were furrowed, then the entire float was

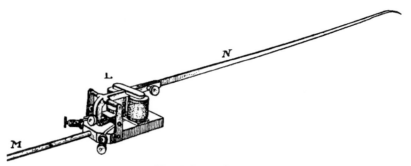

Electric time-marker.

dipped in melted paraffine, the furrows and the paraffine coating practically preventing the usual difficulties arising from capillarity.

Inasmuch as in a number of the experiments as many as five manom-

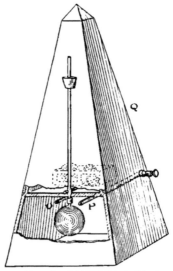

Metronome arranged to make and break the electric circuit at O and P in energizing the time-marker.

eters were recording simultaneously, a device was employed, consisting of a disk of wood, into whose upper surface six strong wires, with ends bent at right angles, were so adjusted as to be free to be swung backward and forward independently. To the free ends were attached long horse-hairs supporting small leaden balls for ballast. Each of the five hairs delicately and steadily maintained the contact of the writing-style of a manometer with the drum, while the sixth was so arranged that, on swinging it away from the drum, it removed all the others simultaneously. The drum-stand was mounted on a crescent-shaped marble slab, and was paralleled by an opening in the table ; to the margins of the latter manometer-holders, arranged by clamps and ratchets so as to be readily adjusted vertically forward and backward, or laterally, giving every facility for adjustment, were fastened. Four drums were used, each ten inches high, adjustable

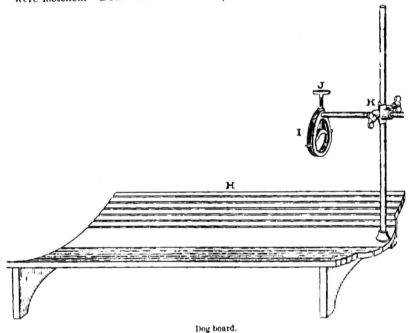

Dog board.

to the same drum-stand, whose motor power was supplied by a double clock-work. These mechanisms were so arranged that by connecting the one or the other with an adjustable governor, any rate of motion, from one revolution in thirty minutes to eighteen revolutions in a minute, might be attained. The rate of motion was indicated by an electrical time-marker. In most of the tracings the events are recorded by means of an electrical signal. The normal salt solution for infusion was maintained at any desired temperature by means of a thermo-regulator.

All the apparatus was constructed by Ulmer & Hoff under my direction.

PROTOCOLS

I

April 25, 1895.—Healthy, adult fox-terrier; weight, nine kilos. Cannula for central pressure in left carotid; for peripheral pressure in right femoral artery; respiration recorded by means of a Marey tambour. Operation upon the head. *First.* Cutting through the skin and clearing away the soft parts from the skull, with rather free hemorrhage, slightly lowered the central pressure, which was closely paralleled by the peripheral; respiration undisturbed; at intervals, usually from thirty to fifty seconds, the central pressure exhibited a marked depressed sweep of a single beat, followed by a few slow beats, then, at an interval of a few seconds, a second drop about equal in extent to the first. This appeared before any surgical procedure had begun. *Second.* The application of the trephine caused a very slight rise in both pressures; no respiratory change. *Third.* On opening the skull no effect was observed. *Fourth.* Cutting away the skull with bone-forceps was followed by a slight rise in both pressures. *Fifth.* Rapidly cutting out the left cerebral hemisphere caused a distinct fall in the central, and simultaneously an abrupt rise in the peripheral pressure, while the respiration showed a deep inspiratory effort, followed by a shallower expiratory, so that in the six following respiratory movements the whole respiratory curve was rapidly raised, then gradually fell again, with the excursions about as before the operation. This temporary rise in the peripheral pressure was followed by a gradual decline, and during this decline the central pressure was as gradually rising, until, at the end of one minute, it had risen to its height at the beginning of the removal of the hemisphere. Following this, the central and peripheral pressures were fairly paralleled with each other. *Sixth.* A similar operation was performed on the right side, with similar results, excepting that when a large piece of skull was suddenly broken away from the meninges there was a fall in both the central and peripheral pressures. The respirations now became gradually less frequent but deeper. *Seventh.* On cutting the dura mater there was a very slight decline in the central pressure and a more considerable decline in the peripheral. *Eighth.* In cutting out the right hemisphere and severing the crus there was a sharp fall in the central pressure, with slowing of the beat, and a rise in the peripheral, the respirations falling from 11 in ten seconds to 6. In thirty seconds the central pressure was steady and showed a tendency to gradually rise, while there was a sharp and continuous decline in the peripheral. For some time afterwards there was a paralleled low pressure in both the central and peripheral, the central showing gradually shorter excursions, the respirations remaining unchanged. *Ninth.* Thrusting a sponge down upon the medulla caused immediate cessation of the heart for fifty-five seconds, when it began beating slowly and powerfully at 24 beats per minute. The peripheral pressure showed a temporary rise during six seconds, then a gradual fall. The central fell at once, close to the abscissa. Respiration failed permanently. *Tenth.* The medulla was then pierced, causing a sudden disappearance of the vagal beats of 24 per minute, which increased at once to 144; then ensued a gradual fall of the central pressure until the death of the animal.

Post-Mortem.—Three hours after death. Small blood-clot in right ventricle. The left ventricle and both auricles empty. The arteries empty. The venous trunks everywhere filled, but especially in the abdomen. The splanchnic veins filled to about the same extent as the somatic veins. The splanchnic arteries contained but little blood.

II

April 30, 1895.—Healthy adult fox-terrier; weight, nine kilos. Central pressure in right femoral, peripheral in left. Operation as in I. Gradual rise, following momentary fall, of the central pressure while clearing away the soft parts over the skull. Free bleeding caused gradual decline in central. Peripheral paralleled. Trephining caused increased frequency of the heart, with slight rise of the pressure, the peripheral following it. Respiration slowed from 21 in ten seconds to 13. While cutting away the left side of the skull with bone forceps there was a considerable decline in the central pressure and less in the peripheral, and the vasomotor curves became more prominent. Cutting out the cerebral hemisphere was followed by a sharp fall in the central, followed in thirty seconds by a gradual rise during fifty seconds to a point nearly as high as before the operation. The peripheral pressure rose while the central fell; then fell during the latter part of the central fall. It fell more rapidly than the central, and again rose more quickly, its curve of rise and fall occupying but twenty-three seconds. The respiratory changes were as in I. Operation on the right hemisphere developed similar phenomena. Digital pressure on the pons excited vagal beats; at the same time complete respiratory arrest, followed by slower respiratory excursions, was observed. Pressure then gradually rose, and, on pithing, a few vagal beats occurred, followed by extremely rapid excursions, gradually diminishing in length, and, while the central pressure gradually declined, the peripheral made rather a sharp rise at the time of pithing, then declined, paralleled with the central. Respirations were permanently arrested.

Immediate Post-Mortem.—Venous trunks everywhere filled; arteries empty and contracted. The splanchnic vessels quite full. The small splanchnic arteries contained considerable blood. The heart in systole, with small clot in each chamber. The pulmonary system anæmic. Time of experiment, one hour.

III

May 2, 1895.—Adult fox-terrier; weight, nine kilos. Central pressure in the right femoral, one hundred and sixty millimetres; peripheral in left, seventy millimetres. While dissecting out the brachial plexus there was a considerable hemorrhage, accompanied by gradual but considerable decline in the central pressure, paralleled by the peripheral. Traction on the brachial plexus with a Wells forceps caused a very marked and rapid rise in the central pressure, displaying vasomotor curves more prominently, the peripheral paralleling it and also displaying such curves. Holding the nerve-trunks with one forceps, they were grasped peripherally by another and roughly torn off, causing a very sharp fall of six millimetres in ten seconds. There were no vasomotor curves during the fall, but they appeared in much longer curves immediately after. The peripheral pressure declined less proportionately. There was no compensatory rise in either pressure after this fall. The respirations were irregular after this manipulation. There was no hemorrhage during this procedure. In a similar way the right brachial plexus was torn out, causing another great decline of ten millimetres in twenty-five seconds in the central, with still longer vasomotor curves than before. The peripheral pressure, however, rose while the central fell, and then, after forty seconds, rapidly declined in about an equal proportion with the central. The respirations were very irregular, sometimes described in long, then in shorter groups of excursions, finally dropping in frequency from 26 in ten seconds to 8 in ten seconds after the lapse of eighty seconds. After sixty seconds there was a gradual but moderate rise in both pressures, the heart excursions growing uniformly and gradually smaller and smaller, the respirations shallower and slower until death. Time of experiment, one hour and twenty minutes.

Immediate Post-Mortem.—Venous trunks full; heart and arteries empty; brain and cord anæmic.

IV

May 6, 1895.—Collie terrier; weight, nine and a half kilos. The central pressure in right femoral artery, peripheral in left. *First.* Crushing of paw with forceps caused a rise of six millimetres in the central, paralleled by the peripheral; respiration augmented. *Second.* Crushed foot extensively, just before corneal reflex was abolished; slight, undulating rise, great rise in peripheral; respirations shallow and rapid. *Third.* Tearing out brachial plexus; fall in central, followed by undulating rise; peripheral, slight rise. *Fourth.* Severely crushed the opposite paw; central rise, peripheral corresponding. *Fifth.* Crushing testes; parallel fall of both pressures. *Sixth.* Cutting of the spermatic cord; slight rise, then fall of both pressures. *Seventh.* Opening abdomen; marked fall of central, peripheral clotted. *Eighth.* Both vagi cut; no change in pressure, but vasomotor curves disappear.

Autopsy.—Heart and arteries empty; veins full, especially the veins at the base of the intestinal loops.

V

May 8, 1895.—Healthy, male fox-terrier, three years old. Central pressure in femoral. Preliminary section of the inferior branches of the right stellate ganglion, by dissection between the scapula and the spinal column, resecting two ribs. Under incomplete anaesthesia crushing of foot caused a very sharp rise, followed by an equally sharp decline of pressure. This was repeated several times. Under full anaesthesia crushing of paws caused rise again. Traction on the brachial plexus caused seven millimetres rise, followed by slight fall. This was repeated with similar results. Cutting the brachial plexus caused a sharp decline of six millimetres. Stretching the proximal ends of the severed nerves caused slight rise, followed by very considerable fall. Incision of abdominal wall followed by slight gradual fall. Opening the abdomen, slight further fall. Pressure upon the abdomen caused marked rise. On release of pressure, sharp fall to a little below the point from which the rise began. Then followed a slight rise to the same height as before the pressure. Respirations were extremely irregular during the stretching of the nerves, less so during the crushing of the paw, and finally grew very shallow and failed before the heart. When the respiratory failure was almost complete the heart's action became very slow, and the excursions of the manometer, extremely long.

Autopsy next day showed left ventricle in systole, the right containing some blood; venous trunks everywhere quite full; the brain anaemic; veins of the brain almost empty. The splanchnic vessels were not more filled than the somatic.

VI

May 9, 1895.—Male dog, mongrel, four years old; weight, twelve kilos; ether; central pressure in carotid, peripheral in femoral. Preliminary attempt at removal of stellate ganglion on left side was abandoned on account of the considerable loss of blood from deep sources. Did not open the chest. The dog was in great shock when the canulæ were inserted into the arteries. The central excursions became gradually smaller. The pressure was so low that further attempt at producing shock was abandoned and intravenous saline solution was administered, causing an immediate rise in the pressure, followed by considerable fall on cessation. The superior and middle splanchnic nerves had been divided before the saline was administered. Respirations gradually grew more shallow, slower, and irregular. On opening the abdomen and manipulating the sympathetic branches, the respiratory efforts were characterized by arrhythmic movements in point of frequency and amplitude of excursions, until finally there was cessation. In all five hundred cubic centimetres of saline were given. Towards the close of the experiment, when the pressure was low, the removal of the intestines, with severe manipulations, caused but little change in the pressure, but great change in the respirations. Respirations suddenly stopped. The heart continued to beat for some time afterwards.

Autopsy was made while the heart was still beating slow and full. The right ventricle was filled with blood, the left empty. The large venous trunks of the limbs full, the arteries empty. The large venous trunks of the abdomen full, the arteries empty.

VII

May 14, 1895.—Dog, mongrel ; weight, nine kilos ; ether. Performed a costo-sternal resection of the anterior chest-wall, tying a double row of ligatures around the ribs and around the sternum, between which sections were made bloodlessly. The flap was then raised up, exposing freely the chest. Through this opening the right stellate ganglion was resected, with the exception of one of the cardiac branches. The left stellate was also only partially resected. But little blood lost. The flap was then replaced, the free ends of the double row of ligatures tied, and the skin sutured. Artificial respiration was substituted by natural. Central pressure in femoral artery, peripheral in carotid. At first there was a gradual decline in the central pressure, with irregular, extremely long excursions, ranging from thirty to forty millimetres in length. During the entire experiment there was a very marked alteration in the character of the heart-beats. At intervals there appeared very decided vagal beats, with their slow and long sweeps, followed by rapid beats beginning with the systole, the blood-pressure rising, then gradually declining to the mean pressure. Very marked vasomotor curves were observed. Such experiments as the crushing of the paw and leg did not cause any definite changes in the pressures. The peripheral closely paralleled the central. After six minutes respiration failed and artificial was supplied. Artificial respiration was attended by a rise in both pressures. The entire experiment was much handicapped by the extraordinary fluctuations in the height of the blood-pressures, the character of the heart-beats, and the recurring failures of respirations.

Autopsy revealed conditions similar to previous ones, though the intestines were more congested than in the preceding case.

VIII

May 15, 1895.—Fox-terrier, two years old; weight, nine kilos; ether. Central pressure in femoral, peripheral in carotid. Beginning central pressure, one hundred and sixty millimetres; peripheral, twenty-six millimetres. *First.* Crushing of the paw was attended by immediate rise of twelve millimetres in the central, eight millimetres in the peripheral pressure, followed by a fall eight millimetres lower than before the experiment, in an interval of twenty seconds, and during forty seconds it again rose to the height it was before the experiment, paralleled by the peripheral. The respirations were a little increased in their amplitude, and in their frequency increased two excursions in ten seconds. Following this their amplitude gradually diminished. *Second.* Crushing of the fore leg was attended by a rise of ten millimetres, followed by a fall six millimetres below its height previous to the experiment, after which, during fifty-five seconds, it rose to the height it was before. After attaining this height it took a second decline during ten seconds, finally reaching a point six millimetres lower than at the beginning of the first experiment. The peripheral closely paralleled the central in its fluctuations. Respirations responded as in the first experiment. *Third.* While dissecting out the axillary space, using blunt dissection mainly, the blood-pressure curves remained unaltered. The respirations were very irregular, both in their amplitude and in their rhythm. After clearing the brachial plexus it was severed with sharp scissors without tension; this was attended by an immediate and gradual parallel decline of nineteen millimetres of both pressures, during an interval of twenty seconds. The respiration remained unchanged. *Fourth.* Crushing of the foot, attended by a rise of twelve millimetres in the central during nine seconds, followed by a decline during twenty seconds to the level before the experiment. The peripheral pressure showed but two millimetres rise in the corresponding time, and after thirty seconds began a gradual rise of ten millimetres in fifteen seconds, independent of the central. Then, independent of the slight central fluctuations, the peripheral pressure maintained this uniform plane two minutes. *Fifth.* Cutting skin of thigh and leg was attended by a rise in both pressures. *Sixth.* Separating periosteum from tibia and fibula caused a gradual rise in both pressures, but more marked in the periphery. *Seventh.* Sawing a number of sections of bone was attended by no change either in blood-pressure or in respiration. *Eighth.*

Crushing of the paw again exhibited the same phenomena as before. *Ninth.* Cutting skin over femur caused a rise of twelve millimetres in ten seconds, followed by a fall to the level before the experiment. The peripheral pressure rose simultaneously ten millimetres. Respiration again altered in rhythm. *Tenth.* The paw was crushed in order to serve as a control before sawing the bone. Usual phenomena followed; then sawing the femur, tibia, and fibula in several places and crushing the bone caused no effect upon respiration or the pressures, and directly afterwards the paw was again crushed as a control, and was attended by a rise as before. *Eleventh.* Opening the abdomen caused no change in pressure. Gently manipulating the intestines caused a slow and gradual decline, from which there was no tendency to return. One hundred cubic centimetres of saline solution temporarily raised the pressure. Following this there was a gradual decline in the pressures, showing no tendency to recover. At this time pressure upon the abdominal walls caused a temporary fall in the femoral, rise in the carotid. Immediate autopsy showed the condition described in previous experiments. During the development of the shock the venous trunks, so far as could be observed, gradually became larger. Respiration failed first.

IX

May 18, 1895.—Brown fox-terrier; weight, nine kilos, three years old (?). Central pressure in femoral, peripheral in carotid. Central pressure, one hundred and seventy millimetres; peripheral, eighteen millimetres. On tying the other carotid the peripheral pressure fell fourteen millimetres; the central rose six millimetres. Crushing the testicle was followed by a fall of thirty millimetres in forty seconds. The pressure then gradually rose, in a period of seventy seconds, to within six millimetres of its height before the experiment. Three subsequent experiments upon this organ were followed by a like result, though in each succeeding one to a less degree. Crushing the paw was followed by a rise, and continual crushing and cutting of the paw by a still further rise of pressure of twelve millimetres. These operations were attended by a considerable hemorrhage; at the same time the animal required ether. These two causes depressed the central pressure twenty-two millimetres during a period of sixty seconds. After this profound depression there was no tendency to a recovery, and cutting the skin, crushing the paw, and otherwise mutilating the extremity caused but little change in the pressures; but the respirations were markedly affected in a manner similar to those described in previous experiments. The peripheral pressure in the carotid throughout this experiment showed but little fluctuation, and did not parallel the central. Before the considerable decline in the blood-pressure the respirations were 12 in ten seconds, and after this decline was well established respirations fell to 8 and gradually failed before the heart.

X

May 22, 1895.—Fox-terrier, three years old (?); weight, eleven kilos; male. Duration of experiment, one hour and twenty-nine minutes. Musculo-cutaneous sterno-costal flap, exposing freely the thorax, through which the stellate ganglia on both sides were completely removed. The flap was closed down, ligatures tied, and skin sutured as before. The animal was allowed to resume natural respirations. Central blood-pressure in femoral, one hundred and twenty millimetres; peripheral in carotid, thirty millimetres. Crushing of the testicle, the paw, the leg, and cutting of the skin extensively caused scarcely an appreciable alteration in either the peripheral or the central. Manipulating the intestines and exposing them to the air caused a gradual decline in the pressure, with shortening of the heart-strokes until death. The heart-beats at the last were very slow, falling from 20 in ten seconds at the beginning of the experiment, to 6 in ten seconds. Respirations failed first.

Immediate Autopsy.—The hind leg, which had been kept elevated and uninjured, was almost totally free from blood; the femoral vein collapsed. Although the fore legs had been subjected to considerable injury in the experiment, they showed full venous trunks. The splanchnic area was apparently not so full as usual. The sinuses of the brain moderately full.

XI

May 22, 1895.—Bull-terrier, four years old (?); weight, nine and a half kilos. Preliminary preparation as in X. Both pressures in the carotid arteries. But little blood lost in the preliminary preparation. Experiment begun at 11.15 and ended at 12.45. Central pressure, one hundred and ninety millimetres; peripheral, eighteen millimetres. Crushing of testicle was attended by slight fall, the respirations slightly altered. Again crushing testicle caused no fall, vasomotor curves becoming more pronounced. Crushing of the paws and of the legs and cutting open the abdomen caused no alteration in the pressure. On opening the abdomen the respirations were altered as described in previous experiments. The blood-pressure gradually declined from one hundred and sixty to one hundred and fifty-two millimetres. Following this there was a gradual decline while the intestines were exposed. Manipulation of intestines in the presence of a slight conjunctival reflex was followed by a rise of ten millimetres during a period of thirty seconds. After further manipulation of the intestines and crushing of the paw the pressure declined to one hundred and thirty-eight millimetres, at which time there began to appear, at rather regular intervals, temporary inhibitions of the heart in single beats, attended by a fall in each such beat of twenty-two millimetres, the up-stroke rising twelve millimetres above the upper pressure curve, making the entire vagal stroke fifty-four millimetres in length. The administration of two hundred cubic centimetres of saline solution caused a temporary rise in the pressure and temporary suspension of the vagal beats. However, after the expiration of seventy-five seconds, the beats reappeared, coming more frequently than before. The respiratory and the heart excursions declined *pari passu* until death. During the injection some air-bubbles inadvertently entered the circulation. While amputating the hip-joint the stopper was dislodged from the canula and a free and depressing hemorrhage occurred. Finally, the femoral artery was opened and the dog was allowed to bleed to death.

Autopsy.—The stellate ganglia were entirely removed; heart and vessels anæmic, depleted of blood; brain and intestines anæmic.

XII

May 23, 1895.—Retriever; weight, fifteen kilos. Duration of experiment, one hour and twenty minutes. Central pressure, one hundred and fifty millimetres; peripheral, forty millimetres; both taken in the femoral. Preliminary resection of eight ribs on the left side, with excision of the stellate ganglion and the chain of sympathetics, was made. Right side intact. Very great difficulty was encountered in the operation. Artificial respirations were readily substituted by natural, but greatly increased in rapidity. *First.* Crushing of left paw caused no change in the pressure. *Second.* Crushing of right paw caused rise of eight millimetres in five seconds after an interval of forty-five seconds since crushing the left paw. During these forty-five seconds there was a gradual decline in the peripheral pressure, deviating from the central in that while the central remained stationary the peripheral was rising. The peripheral made a gently undulating fall of six millimetres. The rise of the central, after the crushing of the right paw, was sustained. *Third.* Again crushing the left paw caused no change in either pressure. *Fourth.* Crushing the right, no change in either. *Fifth.* Crushing the testicles, right and left, —neither caused any change. *Sixth.* Repeated crushing and injury of the legs caused no change in the pressures. In seven minutes of active injury the blood-pressure declined in total but three millimetres. *Seventh.* Opening the abdomen was attended by a small decline in the central pressure. Manipulating the intestines and crushing the same was attended by a very rapid decline of the central and the peripheral pressures; in the central, twenty-four millimetres in eighty seconds; in the peripheral, four millimetres in the same time. The pressures from this point did not rally at any time. Later and during further experiment no amount of injury seemed to cause any change in the pressures after this low point had been reached. Central pressure at the point referred to was one hundred and eighteen millimetres. The respirations grew shallow, and while the central pressure was one hundred and twelve millimetres artificial respiration became necessary. The heart-strokes became shorter and the blood-pressure gradually declined. Respiration failed first. At the beginning of the experiment the number of the heart-beats in ten seconds was 22 ; near the close they were 24 in ten seconds. The dog was finally bled to death by cutting the femoral artery.

Autopsy.—Considerable free blood in the thorax. Heart and vessels anæmic ; brain very anæmic. Intestines show dilatation of the smaller veins.

XIII

May 24, 1895.—Bull-terrier ; weight, eleven kilos. Carotid central pressure, one hundred and thirty millimetres. Time of experiment, one and a half hours. Made preliminary ligation of the right iliac artery with a view to making a bloodless hip-joint amputation, but the collateral circulation was so abundant that a considerable amount of hemorrhage was encountered in the operation. Skin incision for the amputation was attended by a rise in the blood-pressure. The attending hemorrhage caused a very considerable fall. Cutting muscles caused little change. Cutting the bone, none at all. Cutting the sciatic nerve with sharp scissors was attended by a rise of six millimetres in ten seconds, followed by a fall to the same level as before in an equal time. Opening the hip-joint and removing the head of the bone caused no change in the pressure. After this extensive cutting the blood-pressure had fallen but two millimetres. During the cutting of the soft parts the respirations were markedly altered in their rhythm, the effect upon the respiratory movements being strikingly greater than upon the circulation. Subperitoneal ligature of the abdominal aorta caused a rise of but six millimetres in the central pressure. Exposing the sciatic nerve of the other leg and snipping off segments of the proximal stump with sharp scissors caused a very rapid decline in the pressure of sixteen millimetres in forty seconds. The stump was snipped off four times in sixteen seconds. Respirations were not much altered. Extensive cutting of the skin over various parts of the body was attended by a rise in the pressure, followed by a fall, while the respirations were markedly increased in their amplitude, and decreased in frequency from 11 in ten seconds to 9 in ten seconds. Extensive mass cutting of the muscles was attended by a decline of the pressure, but almost inappreciable. Repeated sawing of bone and crushing of head of the femur caused a marked change in the first sawing where the periosteum had not been stripped off. The respirations simultaneously during the sawing were arrhythmic. Following this, five sections of the femur were made with the saw through denuded bone without causing any change in the respiration save slight irregularity and without causing any alterations in the blood-pressure. Opening the abdomen without touching the intestines, and pouring cold water through a funnel into the cavity, caused a fall of twelve millimetres in forty-five seconds. At the same time the respiratory curve rapidly mounted to a higher plane, due probably to increased inspiratory action.

Autopsy.—Right ventricle contained some blood; left ventricle empty; auricles both empty; brain anæmic; large venous trunks full.

XIV

May 26, 1895.—Mongrel; weight, nine kilos. Central pressure in carotid registered one hundred and ten millimetres. Stellate ganglia on both sides resected by the osteo-cutaneous flap method. Crushing the paws and the testicles caused no change in the pressure. Cutting the skin and crushing the brachial plexus, while they caused the usual respiratory change, caused no change in the blood-pressure. Filling the abdomen with cold water caused no change, respirations gradually failed, becoming extremely irregular and arrhythmic, with distinct gasps, followed by very slight respiratory efforts. Hot water turned into abdomen with the cold caused an immediate decline in the blood-pressure, but a reappearance of the respiratory curve after respirations had almost ceased. The decline in the central pressure was not great, however, but was gradually falling. The hind feet in boiling water was attended by a rise in the pressure, and at the same time a rise in the respiratory curve. Crushing of hind limbs was attended by no effect. The respirations after the application of hot water in abdomen were good, but, while the experiment was protracted, they gradually and for a considerable length of time became deeper, and finally, after two hours of experiment, the respirations and the heart-beats still remained fairly good, and the animal was bled to death.

Autopsy showed some of the intestinal coils to have been burned, and in places the tissues coagulated. The ganglia had been completely resected.

XV

May 27, 1895.—Dog; weight, seven kilos. Both stellate ganglia removed. Blood-pressure at beginning of experiment, one hundred and twenty millimetres. Animal in good condition at beginning of experiment. Crushing paws caused no change in the pressure. Placing the hind foot in boiling water for a considerable time produced no change in the blood-pressure. Small abdominal incision made, through which warm water was poured into the abdomen, caused slight rise in pressure and an increase in the heart-strokes ; respiration increased in frequency and in depth. Hot water poured into abdominal cavity caused first a fall, then a gradual rise, in blood-pressure. Respirations were at first greatly accelerated in rhythm with an over-inspiratory tonus. Later the respirations failed before the heart did.

XVI

May 28, 1895.—Fox-terrier ; weight, seven kilos. Both stellates resected as in previous experiment, but at autopsy it was found that a small cardiac branch had not been cut. Blood-pressure, one hundred and fifty millimetres. At one time the anæsthesia was overlooked. The dog became profoundly under its influence, causing a very great fall in blood-pressure of forty millimetres in eighty-five seconds. This fall was recovered in forty seconds on removal of the ether. At the same time respirations were markedly slowed from 6 in ten seconds to 3 in ten seconds. After the blood-pressure had returned to one hundred and forty-five, the abdominal aorta was ligated just above its bifurcation, which was attended by a rise of six millimetres. Hip-joint amputation was performed. Cutting of skin caused no change in blood-pressure, but the usual change in respiration. Quickly cutting the sciatic nerve with sharp scissors was attended by a fall of eight millimetres ; respiration unchanged. Sawing the femur caused no appreciable change. Boiling water poured on the intestines caused a temporary rise ; then gradually the pressures declined, while the respiratory excursions were markedly increased in their amplitude and in their frequency from 6 in ten seconds to 8 in ten seconds. Autopsy showed the heart empty. Large venous trunks full as in previous experiments.

XVII

February 29, 1896.—Dog ; weight, eight kilos. The animal had been subjected to laryngeal experiment, and had received intravenous injection of one one-hundredth grain of atropine. Dog with his body inclined head downward. Boiling water was poured into the abdominal cavity. Respirations were slowed during three excursions, then very much accelerated with extreme amplitude. Blood-pressure was rapidly lowered, with perhaps a slight preliminary rise. After the blood-pressure had recovered itself, several loops of intestines were withdrawn from the cavity and placed in boiling water, attended by a rapid rise of the blood-pressure, followed soon by a fall.

Autopsy.—The scalded intestines were colorless. The loops that had been exposed to warm water extremely congested. The brain was much congested.

XVIII

Large dog, under surgical anæsthesia. *First.* A hæmostatic forceps was made to exert a pressure within the larynx, which was attended by arrest of respiration, and with an abrupt fall of blood-pressure almost to the abscissa. *Second.* Applied four per cent. solution of cocaine locally ; then repeated a like experiment, and in this case there was no change in either the blood-pressure or in the respiration. *Third.* With the dog inclined head downward, boiling water was poured into the abdominal cavity through a funnel. Respirations became deeper and more rapid ; the blood-pressure was but little altered.

XIX

First.—Laryngeal experiments similar to the preceding, applying considerable quantity of cocaine with practically the same results, were first made. *Second.* The abdomen was opened, all the intestines, the stomach, and the spleen were withdrawn and exposed upon the outer abdominal wall. No change in the blood-pressure was noted. *Third.* On manipulation of intestines there was a gradual rise. When manipulations ceased there was a fall to about the same level at which the rise began. Rough and extensive manipulation quickly caused some rise, and was followed by a fall to the original level, when manipulation ceased. During this time the intestines became intensely congested.

XX

One-fortieth grain of cocaine was administered to a fifteen-kilo dog, and was attended by an acceleration of the respiration and a rise in the blood-pressure, followed by immediate fall to the level before. Previous to this injection, a control had been made, in which intra-laryngeal manipulation caused arrest of respiration and inhibition of the heart. After the cocaine injection, on repeating the experiment, the respirations were arrested as before, but the blood-pressure remained unaltered. Then applying four per cent. solution of cocaine locally on the laryngeal mucosa, and repeating severe manipulations, no change on either the blood-pressure, the heart-beats, or the respiration was caused. Abdomen opened, intestines extensively and rudely manipulated ; no change in blood-pressure, but deepened respiratory rhythm. No amount of manipulation and injury of intestines caused any change in the blood-pressure. Crushing of the testicle, twisting and tearing the sciatic nerve, likewise caused no change. Intestines were greatly congested.

XXI

May 25, 1896.—Dog ; weight, eight kilos. *First.* A control experiment on the larynx by intra-laryngeal manipulations exhibited a heart inhibition and respiratory arrest as before. *Second.* Injection of one two-hundredth grain of atropine caused very great rise in blood-pressure, with shorter excursions and shorter beats. Intra-laryngeal manipulation was attended by a rise in the pressure without altering the character of the beats. The respiratory alterations were possibly less marked than in the control experiment. Noted that the salines applied to the chest caused increased respiratory rhythm. On manipulation of intestines, though less severely than in XX, there was a fall in blood-pressure, without any preliminary rise. The intestines became extremely red and congested.

XXII

Central pressure in carotid. *First.* On manipulation of intestines there was a very marked disturbance in the rhythm and the amplitude of the respirations, so much so that it was almost equivalent to an arrest. After exposure of the intestines, they became much congested and the blood-pressure suffered a slow decline. The stomach especially was markedly congested. *Second.* All the intestines were removed and rudely manipulated, causing further fall in the blood-pressure and very great respiratory changes, as above. *Third.* Hypodermic injection of one-twelfth grain of cocaine. *Fourth.* Contact with peritoneum still caused the respiratory alterations as before, but the blood-pressure was unaltered.

XXIII

June 11, 1896.—Experiment similar to preceding. *First.* Slight contact with the visceral peritoneum caused immediate respiratory changes, as described in XXII. This was true of the parietal peritoneum everywhere, as well as of the visceral. *Second.* A two per cent. solution of cocaine applied over a given area of peritoneum. *Third.* Manipulations of the cocainized area produced no respiratory changes. *Fourth.* Solution of sulphate of magnesium from the pressure-bottle applied to the exposed vagi caused an increase in the frequency and in the depth of the respirations.

XXIV

June 12, 1896.—Experiment as in XXIII. *First.* Manipulation of the mesentery, no effect. *Second.* Manipulation of the intestine showed some change in the respiratory rhythm, not so marked as in the preceding; blood-pressure slowly fell. *Third.* Compressing the abdominal aorta caused temporary rise, followed by gradual fall. *Fourth.* Compressing the abdomen with open hand caused a rise,—higher rise than clamping the aorta. Intestines became congested upon exposure. *Fifth.* Larynx exposed; intra-laryngeal manipulation caused arrest of respiration and inhibition of the heart. After the application of cocaine the reapplication of the laryngeal irritation produced no changes.

3

XXV

June 13, 1896.—*First.* Severe manipulation of the omentum was attended by no change. *Second.* Similar manipulation of intestine was followed by a marked respiratory change, more marked than the preceding, with a gradual fall of blood-pressure. The nearer the manipulations were made to the diaphragm the more marked were the respiratory changes. Parietal as well as visceral contact was attended by these phenomena. *Third.* Made a retro-peritoneal exposure of the abdominal aorta and clamped the same, supposedly above the cœliac axis; then began manipulations of intestines, and fall followed as before.

Autopsy proved the supposition wrong; the cœliac axis had not been included. Clamping the aorta caused a rise of pressure. The intestines became livid. The veins, especially those at the base of the intestines, were greatly dilated.

XXVI

June 18, 1896.—*First.* Aorta retro-peritoneally clamped above the cœliac axis, between the pillars of the diaphragm. Blood-pressure rose, and the dog was shaken by tremors in all parts of his body. *Second.* On manipulation of intestines there was a fall of ten millimetres during three seconds, and in five seconds more the fall was recovered, followed by a gradual rise in seventy seconds to twenty millimetres higher than before the experiment was begun. This high level was sustained. After some time there was a repetition of the vagal beats, which supervened soon after the clamping of the aorta. *Third.* On removal of the clamp there was an immediate staggering fall of the blood-pressure, followed by an incomplete compensatory gradual rise. The splanchnic vessels all became engorged. The intestinal coils up to the time of the unclamping of the aorta were white and contracted.

XXVII

After preliminary clamping of the aorta, when ready for the experiment, the cork was dislodged from the canula, causing frightful hemorrhage and rapidly sinking blood-pressure. Respirations were deep and rapid during this time. The canula was then placed in the opposite carotid, the intestines manipulated, attended by a decline in pressure and flushing of the vessels.

Autopsy.—The forceps had slipped and was grasping but about three-fourths of the vessel. The large vessels were everywhere contracted in the somatic area, the arteries as well as the veins; splanchnic area congested; right heart engorged, left almost empty; lungs collapsed and their vessels empty.

XXVIII

First. The superior mesentery area was clamped subperitoneally. Some hemorrhage was encountered in the operation. The right renal was also ligated. *Second.* On opening the peritoneum the intestines were found somewhat congested. Handling the intestines was attended by a wavering line, without any decline in the pressure. Manipulations were continued actively for one hundred seconds, at the end of which time the blood-pressure was four millimetres higher than at the beginning. *Third.* Smartly slapping the intestines with large scissors was attended by a rise of four millimetres in ten seconds; on cessation there followed a slight decline of pressure. Rapid and continuous slapping was continued for one minute, when the blood-pressure was at the same height as it was before the slapping. Then, allowing the pressure to become steady, a repetition of the slapping with the handle of a scalpel was attended by a rise of eight millimetres in fifteen seconds. On cessation of the slapping there was a slight decline of the pressure. *Fourth.* On further manipulation of the intestines they became paler. *Fifth.* The abdominal aorta was clamped and the hind extremities were extensively incised; attended by no appreciable change. *Sixth.* Hip-joint amputation made on both sides caused a rise in pressure. *Seventh.* Grasping the sciatic nerve with forceps and making traction caused a rise. The heart excursions gradually grew smaller, respirations more shallow. Temperature at close of experiment $100\frac{1}{2}°$ F. *Eighth.* On removal of clamp from mesenteric artery the vessels became extremely dilated and the intestines red. No change was observed in the vessels of the omentum. The tissues of the somatic area in their circulatory aspect were just the opposite to those of the splanchnic area.

XXIX

First. Superior mesentery clamped almost bloodlessly. *Second.* On manipulation there was an almost immediate fall, followed by an immediate rise to its former level. Respirations altered as before. When first picked up the intestines were somewhat congested. Manipulations paled them. *Third.* Striking intestines with scissors caused rise in the pressure. On unclamping the artery there was an immediate fall of fifty-four millimetres in thirteen seconds. The compensatory rise was but eighteen millimetres; then, after allowing sufficient time for compensation, further manipulation, though only slight, was attended by a fall of forty-six millimetres in forty-five seconds. Free hemorrhage from the carotid soon caused the death of the animal. Accelerated breathing immediately followed the beginning of the bleeding.

XXX

June 24, 1896.—Spaniel bitch, with enlarged thyroids, vessels of neck numerous and bluish. Hemorrhage unusually profuse from smaller vessels, not only from neck but elsewhere. Carotid arteries abnormally large in proportion to size of dog; other arteries not so markedly enlarged. Blood-pressure one hundred and twenty millimetres at beginning, ninety at end of experiment. Temperature at end of experiment 99½° F. *First.* Clamped retro-peritoneally the superior mesenteric artery, causing a rise of ten millimetres. On opening the abdomen the intestines were found

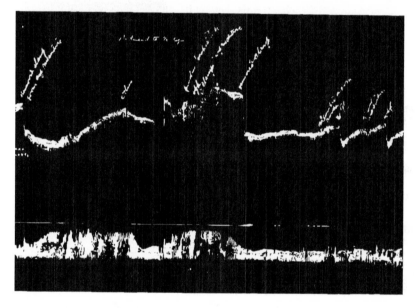

CLAMPING THE SPLANCHNIC VESSELS AND ABDOMINAL AORTA.—The upper tracing represents the blood-pressure; the lower, respirations. Showing effect of clamping and unclamping the superior mesenteric artery and the abdominal aorta, the latter below the splanchnic branches. Note the effect of administering ether.

abnormally congested. While manipulating them they became pale and contracted. Blood-pressure rose gradually twelve millimetres in eighty-four seconds. Respirations were altered as usual, irregular, prominent excursions. After cessation of manipulation the blood-pressure gradually fell to the level before occupied. *Second.* Considerable variation was noted while going in to remove the clamp. *Third.* Removal of the clamp. There was a considerable fall in the pressure, followed by a partial recovery. *Fourth.* In clamping the cœliac axis alone there was a rise in the pressure. Manipulations of intestines caused additional rise in pressure.

Fifth. In removing clamp from cœliac axis there was a very considerable immediate fall of pressure. On clamping the cœliac axis the blood-pressure rose thirty millimetres. Removal of clamp from the cœliac axis caused a fall of twenty-four millimetres, which was followed by a gradual recovery of fourteen millimetres. Clamping of the abdominal aorta was attended by a rise of four millimetres. Unclamping resulted in a fall of sixteen millimetres, followed by complete and rapid compensation. The blood was tardy in coagulating.

XXXI

Large, long-haired, black dog. Experiment as in XXX. Thyroid somewhat enlarged; blood-vessels prominent. *First.* After clamping the aorta above the mesenteric artery and partially clamping the cœliac axis, the intestines were found much congested. *Second.* Their manipulation was attended by a rise in blood-pressure, but striking them caused an abrupt rise. The intestines did not become so pale or so much contracted as in XXX. *Third.* On ceasing manipulations, blood-pressure fell to about the same level as before. *Fourth.* Stretching the rectum produced no effect. *Fifth.* Repeating intestinal experiment, the same result was again obtained. *Sixth.* Pouring ether into tracheal tube caused sudden cessation of respiration; heart continued to beat; blood-pressure fell first rapidly, then gradually to the base-line.

XXXII

July 8, 1896.—Small terrier. Bloodless sub-peritoneal clamping of both renal arteries, superior mesenteric arteries, and the arteries of the cœliac axis. *First.* Withdrawal of the intestines and manipulations, attended by a marked and instant fall, then at once a rise to the same height as before the manipulations. *Second.* Repeated the first experiment with similar results. *Third.* Sharply striking the intestines was attended by a rise in pressure. On ceasing, the pressure fell to its previous level. *Fourth.* On unclamping the vessels, a drop of eighteen millimetres followed, and but little compensation occurred afterwards. *Fifth.* Reclamped the vessels; sharp and sudden rise, which was sustained. *Sixth.* Unclamped; fall as before. *Seventh.* After the last fall, further manipulations caused an additional fall. *Eighth.* Clamping abdominal aorta before the splanchnic caused but slight rise of pressure; unclamping, the corresponding slight fall. *Ninth.* Pressure on the abdomen; very great rise, remaining as long as the pressure. Vessels became pale each time when they were smartly slapped. Each time the vessels were unclamped breathing was augmented.

XXXIII

Clamped the cœliac, mesenteric, and left renal arteries. Accidentally punctured the attachment of the diaphragm on the right side. Respiration suddenly stopped, blood-pressure fell almost to base-line. On artificial respiration both recovered. Respiration became very much exaggerated, then another cessation, artificial respiration, then recovery. This failure, followed by hyperpnœa, occurred a number of times. The left fore leg, then the left hind leg, and, finally, the whole body became tetanized. The dog was finally allowed to die in his own way. The heart beat on for some time after the respirations ceased.

XXXIV

Large dog. Splanchnic arteries clamped ; stellate ganglia on both sides removed. Intestines exposed. *First.* Manipulations, severe or slight, were attended by very slight decline in the blood-pressure in several instances, but never by a rise. The pressure curves for a long time were quite irregular. *Second.* Removed clamps ; staggering fall followed, and soon death occurred. The enormous fall after a time but slightly readjusted itself. Later manipulations caused further fall, but not marked.

XXXV

First. Clamped splanchnic arteries. *Second.* Osteo-cutaneous resection of the thorax. *Third.* Intestinal irritation control ; slight fall, then rise. *Fourth.* Severed the cardiac branches of both stellates ; repeated third experiment ; no change except a very slight decline. *Fifth.* Splanchnic unclamped ; great and permanent fall. Autopsy verified the technique. Temperature at death, $105\frac{1}{5}°$ F.

XXXVI

July 16, 1896.—Small white dog. *First.* Manipulation of omentum caused a considerable fall, then a rise above normal. Repeated manipulation three times with like results. *Second.* Manipulation of intestines caused a rapid fall at first, with secondary partial recovery, then gradual fall to a low level ; no compensatory rise. Intestines became much congested. *Third.* Squeezing a coil of the congested intestine was attended by a slight rise. *Fourth.* Pressure on the abdomen caused a rise. *Fifth.* Hot water introduced into abdominal cavity produced a rise. *Sixth.* Warm water, a fall. *Seventh.* The same.

SPLANCHNIC VESSELS CLAMPED, THEN INTESTINAL MANIPULATION AND CLAMPING AORTA BELOW.— Upper tracing, blood-pressure. Lower, respirations. Splanchnic arteries clamped and stellate ganglia removed previous to taking this tracing. 1. No effect noted on severe injury of the intestines while freely exposed. 2. Clamping abdominal aorta; note sustained rise without showing usual compensatory fall when similar procedures are done with open arteries and intact stellates.

XXXVII

July 16, 1896.—Splanchnic vessels clamped and cardiac branches of stellate cut by resecting ribs on both sides. Splendid regular tracings of both respiration and blood-pressure. *First.* Omentum severely manipulated, *nil. Second.* Intestines removed and manipulated, *nil.* But on returning them there was an immediate fall. *Third.* The same. *Fourth.* When gentle pressure was made in the abdominal cavity, but not on the intestinal wall, a fall, then a rise. When vena cava was compressed, fall. *Fifth.* Severely slapping the intestines, no effect beyond several individual strokes of the heart. *Sixth.* Abdominal aorta clamped, an immediate considerable rise, and so remained. *Seventh.* Unclamped, fell to the previous point before the artery was clamped, then rose to the same point. *Eighth.* Removal of clamps from the splanchnics, followed by rapid fall and rather sudden death. Autopsy; the technique was shown to have been completely performed.

XXXVIII

July 18, 1896.—Splanchnic vessels clamped; cardiac branch of stellate cut, and during the latter operation a vein of considerable size was torn, causing hemorrhage. Dog was weak before experiment was begun. *First.*

Manipulating of omentum, *nil*. *Second*. Manipulation of intestines, *nil*. Dog gradually failed; respiration ceasing first.

XXXIX

July 23, 1896.—Technique as in XXXVIII. But few drops of blood lost in preliminary preparation. Respiration curve regular; blood-pressure varied extremely. *First*. Manipulation of omentum, *nil*. *Second*. Of intestines, *nil*. *Third*. Rough handling, *nil;* later a very great fall. Anæsthesia at this time was deep. *Fourth*. Manipulation of intestine, *nil*. *Fifth*. Clamping of aorta; a rise, which remained without the ordinary compensatory fall. *Sixth*. Unclamped the aorta; fall, followed by a compensatory rise.

XL

Splanchnic vessels and cardiac branches of stellate treated as before. Dog very small and poorly nourished. Died soon after the manometer was applied. Respiration failed first; heart continued to beat for some time afterwards.

XLI

Very small and old dog, suffering from mange. Died by failure of respiration after the cardiac branches of the stellate were cut. Pulse remained regular and slow to the end of the experiment.

XLII

July 30, 1896.—Cocaine experiment. Twenty-pound dog. Carotid pressure. Time of experiment, one hour and fifteen minutes. *First*. Three minims of a two per cent. solution of cocaine injected into jugular vein was attended by an immediate rise of eight millimetres in central pressure and a decrease of the respiratory curve to one-third its amplitude during thirty-three excursions. An additional injection of two minims was attended by a rise in the blood-pressure, with diminished respiratory curves about as before. Very marked vasomotor curves, with some irregularity, were noted for some time after this injection. *Second*. On cutting through the abdominal wall there was a considerable fall in the blood-pressure, followed by a compensatory rise six millimetres higher than before. *Third*. Removal of intestines and their manipulation caused some fall in the pressure. On their return, blood-pressure rose to its previous height. The intestines became very dark, the veins gradually dilated, arteries very small. Heart-strokes became smaller, respiration irregular, and dog died.

XLIII

August 1, 1896.—Shepherd dog; weight, sixteen pounds. *First.* One-twelfth grain of cocaine was injected into the jugular vein. Vasomotor curves appeared soon after; no respiratory changes noted. *Second.* Manipulation of intestines for a considerable length of time caused no change in the blood-pressure; respirations irregular. The beginning of this manipulation was attended by a rise of eight millimetres. *Third.* More severe manipulations were attended by an irregularity in the heart-stroke, but no fall in the pressure. Respiration suddenly stopped during this manipulation, and artificial respiration was supplied, but it was noted afterwards that the clamp of the tube leading to the pressure-bottle had slipped. It was not known when it came off, whether during or before the experiment. In two previous experiments the removal of this clamp caused sudden death.

XLIV

August 2, 1896.—Time of experiment, one hour and twenty-five minutes. One-twelfth grain of cocaine injected into the jugular vein caused temporary rise in blood-pressure. Manipulation of intestines caused a rise in the pressure, followed by a compensatory fall to its former level. Each successive manipulation caused a rise, then a fall to the original level. At the end of the experiment the blood-pressure was as high as at the beginning. The intestines became red; the large vessels became smaller. The respirations in each manipulation were very much augmented in rhythm and rapidity, and, though the reflexes were absent at the beginning, they became apparent at once during manipulation. Carotid clotting occurred at short intervals. Artery forceps had been used to clamp the vessel. The other carotid was substituted and the weak compression-clamp used, and in this case there was no clotting. Finally, injected three and one-half drachms of two per cent. solution of cocaine into jugular vein, and the dog died of slight convulsions and failure of respirations. The heart beat slowly to the last. Splanchnic arteries became almost obliterated.

XLV

August 4, 1896.—Bull-dog; weight, thirty-six pounds. Time of experiment, three hours and ten minutes. Carotid pressure, one hundred and ten millimetres in the beginning; highest, one hundred and twenty millimetres. *First.* One-twelfth grain of cocaine injected into the jugular. On the injection of cocaine there was a rise of fourteen millimetres; after the lapse of a few minutes there was a more gradual fall; but there was a permanent gain of six millimetres in the height of the pressure. *Second.* Manipulation of the mesentery was attended by a rise of eight millimetres. *Third.* Manipulation of intestines was attended by a rapid fall, followed by a gradual rise to the same level as before. The curves now became irregular, and when finally steady there was a gain of ten millimetres in height over what it was before the cocaine was originally injected. *Fourth.* Repeated manipulation was attended by a similar fall, followed by a similar gradual rise to its former level. The pressure curve was extremely irregular. The respirations were altered as before. Further continuous manipulation was attended by similar phenomena. The intestines were manipulated for half an hour, at the end of which time the pressure had lost but two millimetres. After a period of rest, the intestines were returned and manipulation repeated, finally leaving the blood-pressure, at the end of seventy minutes, higher than it was before the first injection of cocaine. During the experiment, when the dog began struggling, the conjunctival reflexes did not respond. This was observed several times, misleading the anæsthetizer. The splanchnic vessels were not dilated, while the intestines were reddish in color. The vessels were quite small. The transparent areas of the mesentery were injected with blood. After this part of the experiment was concluded, the dog was subjected to all kinds of operations for a period of three hours more, and was finally killed by injecting one drachm of a two per cent. solution of cocaine. There was no clotting of the blood during the entire time. Among the operations performed were double hip-joint amputation, amputation of the fore legs, the application of a Murphy's button, and nephrectomy.

XLVI

August 6, 1896.—Black dog; weight, six kilos. Duration of experiment, fifty-five minutes. Beginning carotid pressure, one hundred and twenty millimetres. Injection of cocaine was immediately followed by a rise of thirty-six millimetres, with decrease of the amplitude of respiration. Severe and continued manipulation of the intestine was attended by no change in the blood-pressure or of the respiration. During the thirty-five minutes of abdominal experiment the blood-pressure remained at practically the same mean level. At the end of this time two minims of the cocaine solution were injected. On clamping the aorta the dog died. Heart failed first.

XLVII

Dog; weight, six kilos. After the injection of the cocaine, the same phenomena as in the preceding were observed. The intestines were freely exposed, and after a long-continued manipulation the blood-pressure remained steady at one hundred and forty-three millimetres. The rhythm of the respirations was somewhat disturbed by these manipulations. After long and continuous manipulations and exposures of various kinds, the blood-pressure, having during this time undergone many sudden changes, finally registered as high as when the experiment was begun.

XLVIII

Atropine experiment. Dog; weight, fifteen kilos. Blood-pressure, one hundred and fifty millimetres at beginning. *First.* Injected one one-hundredth grain of atropine into jugular at two intervals; attended by a fall of two millimetres, followed by a gradual rise of eight millimetres and a compensatory fall, finally taking a level eight millimetres below its previous point. *Second.* Manipulation of the mesentery caused a rise in pressure. *Third.* Manipulation of the intestines was attended by an immediate appearance of vasomotor curves and a fall of ten millimetres, afterwards a rise of fourteen millimetres, and later took a level six millimetres below the control. The curve was extremely irregular, and at the end of the first drum it was precisely one hundred and fifty millimetres. At the end of the experiment—two hours—the pressure was steady at one hundred and forty-five millimetres. The dog was afterwards subjected to numerous operations, as in the preceding cocaine experiment, for two hours more. Finally killed by over-anæsthesia.

XLIX

A badly nourished yellow dog, with large thyroid glands. An initial pressure of one hundred and twenty-five millimetres. Was given one one-hundredth grain of atropine. The heart excursions were rather short. Intestinal manipulations were made similar to preceding in character and extent, and at the close of the first drum the pressure was as high as it was in the beginning, after a duration of thirty minutes. In the second half-hour the respiratory curve was so extremely irregular and clotting so frequent that the experiment was handicapped to such an extent as to render the tracings useless.

L

Injection of one one-hundredth grain of atropine, subjecting the animal to a similar experiment. There was also repeated clotting, but manipulations did not cause a lowering of the blood-pressure.

LI

Dog; weight, six kilos. *First.* Laryngeal control; there was respiratory arrest, and cardiac inhibition on making intra-laryngeal manipulation. *Second.* Applied two per cent. solution of cocaine to laryngeal mucosa. On repeating the manipulation the above phenomena were abolished. *Third.* Touching the pharynx through the larynx caused reflex swallowing. *Fourth.* Cocainized the same. *Fifth.* Touching in like manner, no reflex swallowing. *Sixth.* Cutting the scrotum in the ordinary manner, fall in blood-pressure and respiratory changes. *Seventh.* Compressed testes with forceps, considerable fall. *Eighth.* Traction on spermatic cord, fall. *Ninth.* Other testicle and cord showed similar phenomena. *Tenth.* Opened the abdomen and exposed the intestines without manipulation; no change. The vessels seemed congested. *Eleventh.* Intestinal manipulation was attended by the usual great fall. *Twelfth.* Sudden crushing of the testicle was followed by but little change in the blood-pressure.

LII

December 7, 1896.—Large dog. Testicle and larynx. Testicles quite small. *First.* Laryngeal control; respiration arrested; blood-pressure fell almost to zero. *Second.* One one-hundred-and-fiftieth grain atropine in the jugular. Second experiment repeated; no fall in blood-pressure, but the heart-beats were accelerated; respiration arrested as before. *Third.* Testicle manipulated; incision made through the skin and tunica vaginalis; marked fall. *Fourth.* Fifteen minims of a two per cent. solution of cocaine injected into the testicle. *Fifth.* Further manipulation after a control; no change. *Sixth.* The other testicle cocainized by hypodermic injections into the tunica vaginalis and along the spermatic cord. Severe manipulations; no change in the pressure. *Seventh.* A loop of intestine removed and cocainized locally to determine whether cocainization would prevent the contraction of the blood-vessels. *Eighth.* Withdrew the intestines and exposed them to the air; dilatation of the vessels and fall in the blood-pressure followed. *Ninth.* A dose of atropine sufficient to inhibit the vagus was given, and still there was a fall in blood-pressure on manipulating the intestines.

LIII

December 8, 1896.—Wolf-dog; weight, ten kilos. Testicle and larynx. *First.* Laryngeal control by means of intra-laryngeal pressure, producing respiratory arrest and inhibition of the heart. *Second.* A two per cent. solution of cocaine applied to the laryngeal mucosa. Immediate repetition of the above control was attended by no alterations in either respiration or the heart-beats. Extending the finger upward into the pharynx caused swallowing. The application of cocaine solution immediately prevented swallowing on a repetition of like manipulation. *Third* Testicular control. By gently cutting the skin, including the tunica vaginalis, there was a marked fall. *Fourth.* Ten drops of two per cent. solution of cocaine injected into the tunica vaginalis and into the spermatic cord; immediate rise of the blood-pressure, but at the moment the skin was picked up and the needle inserted there was a fall. *Fifth.* Further cutting and severe manipulation was not attended by a fall. *Sixth.* The other testicle likewise cocainized and likewise treated; no fall. *Seventh.* Spinal cord cut at the upper dorsal vertebræ. Very great fall of pressure, followed by a slow and incomplete compensatory rise. *Eighth.* Intestines removed, exposed, and gently manipulated, attended by some additional fall. Faradic stimulation of the distal segment of the cord caused a rise. *Ninth.* Stimulation of the omentum caused a rise. *Tenth.* Elevating the foot of the dog-board, a rise. *Eleventh.* Clamping aorta, a rise. The splanchnic area was now extremely congested; the heart-strokes became faster and shorter until the death of the animal.

LIV

December 9, 1896.—Hunchback cur. Low blood-pressure, enlarged thyroids, carotids disproportionately large in comparison with the abdominal aorta. *First.* Laryngeal observations as in LIII. *Second.* Sciatic nerve exposed. Faradic stimulation; rise of blood-pressure. Intended to make further observations on the use of cocaine, but the blood-pressure was so low, the dog showing so many signs of exhaustion, that this was abandoned. *Third.* On removal of intestines there was a great fall. *Fourth.* On stimulation of the sciatic respirations were increased in their frequency and augmented in their amplitude. Digital contact in various parts of the peritoneum caused striking respiratory alterations. *Fifth.* A few drops of salt solution happening to fall upon the open larynx caused irregular catching inspirations. *Sixth.* Clamping the aorta above the cœliac axis produced a rise. No compensatory fall followed. *Seventh.* Unclamping the aorta and reclamping just above its bifurcation produced rise. Compensation then followed. *Eighth.* Application of cocaine on the mesentery did not prevent contraction of muscular coats of intestines on stimulation. The heart-strokes gradually grew shorter and shorter, the blood-pressure declined, and the animal died.

LV

December 10, 1896.—Large dog. Duration of the experiment, one and a half hours. *First.* Cut through the floor of the mouth and exposed the pharynx. Contused and manipulated the epiglottis, upper surface of the pharynx, and rima glottidis. No respiratory or circulatory changes were noted. *Second.* Intralaryngeal contact, though slight, exhibited the respiratory and circulatory changes, as in LIII. *Third.* Local application of one-half per cent. solution of cocaine upon the laryngeal mucosa prevented the above phenomena on repeating like manipulation. *Fourth.* Various parts of the trachea and cricoid manipulated severely. No alterations in the blood-pressure or respiration. *Fifth.* Testicular test followed by marked fall of blood-pressure; respiration unchanged. *Sixth.* A few drops of one per cent. solution of cocaine was injected into the spermatic cord of other testicle. On manipulation of this organ there was very slight fall, much less than in No. 5. *Seventh.* Opening abdomen and striking intestines caused great fall. Respiratory changes as before. *Eighth.* Sciatic nerve exposed and severed. Proximal end faradized; marked rise in pressure; respiratory frequency and amplitude increased. *Ninth.* Cocainized opposite sciatic nerve. Faradization of this nerve then caused considerable fall, and no respiratory changes were noted. In No. 8 the heart-beats were very irregular, in this experiment regular. *Tenth.* Extirpation of an eye and rude manipulation and bruising of the socket; no manifest effects.

Temperature.—When first under ether at 2.20 P.M., 38° C.; at 2.50 P.M., 39° C.; at 3.03 P.M., 39.5° C.; at 3.10 P.M., 39° C.; at 3.17 P.M., 38.5° C.; at 3.40 P.M., 37° C.; death at 3.50 P.M.

LVI

December 12, 1896.—Dog; weight, five kilos. Time of experiment, two hours. *First.* Cocainized one spermatic cord with one-half per cent. solution of cocaine at some distance above the testicle, within the inguinal canal. *Second.* Exposed both organs. *Third.* Squeezing the cocainized side, but little fall. *Fourth.* Squeezing the other side, great fall. *Fifth.* Exposed sciatic nerve; severed it. On picking up proximal end and inserting needle before cocaine was injected there was a rise. *Sixth.* After the cocaine had been injected, picked up the proximal end with forceps and applied faradic current; marked rise in pressure. Suspected that the traction on the nerve may have caused the rise. Faradization was then done without traction, followed by no rise in pressure. No respiratory changes. *Seventh.* Repeated No. 6 with like results. *Eighth.* Brachial plexus then treated in similar manner, with like results.

LVII

Brown cat. *First.* Exposed sciatic. Faradization attended by rise in blood-pressure. The sciatic sheath cocainized. Faradization above the cocainized point caused contraction of the distal muscles. *Second.* Laryngeal stimulation as in the preceding; respiratory arrest. Animal died rather suddenly. Heart excursions became uniformly smaller until death. Respiration failed first.

LVIII

December 16, 1896.—Spaniel bitch; weight, three and five-tenths kilos. Duration of experiment, one hour. *First.* Manipulation of one ovary attended by a rise in blood-pressure. During this manipulation breathing became irregular. *Second.* The other ovary manifested the same phenomena. *Third.* Digital contact with various portions of the parietal peritoneum was attended by irregular respiratory rhythm. *Fourth.* Animal was killed by forcing air into the left jugular vein.

LIX

December 22, 1896.—Spaniel bitch; weight, six kilos. Very cross; had been nursing puppies. Duration of experiment, one and a half hours. High blood-pressure. *First.* Manipulation of ovaries caused slight rise in blood-pressure; not so marked as in LVIII. Manipulation of tubes and uterus; no perceptible change. *Second.* Forcible dilatation of sphincter ani; slight increase in respiratory rhythm. *Third.* Animal killed by pouring chloroform into the tracheal canula and clamping the tube. Sudden cessation of the heart and of the respiration.

LX

December 23, 1896.—(Marked LIX on tracing.) Genito-urinary experiment; bitch; rather low blood-pressure. Manipulating a number of times the ovaries, tubes, and uterus, imitating operative proceedings, attended in each instance by rise of pressure. Believed also to be attended by change in respiration, but not quite certain as to whether or not the mechanical element could be eliminated. One apparent exception in the case, traction on both tubes, attended by immediate and considerable fall of blood-pressure. On further investigation, it was more than probably caused in a mechanical way by impeding the circulation in the large venous trunks. The temperature began gradually to decline from the beginning of the experiment, from 39° to 37° C. at death. The most marked rise in the pressure occurred on manipulation of the uterus, combined with considerable traction. Manipulation of the bladder was attended by no appreciable change.

LXI

December 24, 1896.—Bitch; weight, seven kilos; genito-urinary experiment. Duration of experiment, one and a half hours. *First.* On manipulation of uterus, ovaries, and tubes, there was a rise of blood-pressure, in degree in above order; injuring the vagina produced a smaller rise. *Second.* Exposing the intestines to the air a great congestion was seen to develop and a fall of pressure occurred. After a very considerable fall, with the blood-pressure almost at zero, intravenous infusion was made. One thousand cubic centimetres of plain water at 50° C. caused at first a very marked rise in blood-pressure, but rapidly fell to about the same level as before. The strokes became longer. Five hundred cubic centimetres additional directly and continuously infused had but little additional effect. The veins of the splanchnic area were extremely dilated. On making a further dissection the bleeding was unusually free, and blood showed but little tendency to cease or to clot. The blood, of course, was markedly pale. The temperature fell gradually from 38° C. at the beginning to 33.5° C. at the time of death. Although this very large amount of water much warmer than the blood was infused, the temperature rose but one-fifth degree. The temperature in all these cases was taken per rectum. The amount of water infused was about twice the amount of blood calculated to have been in the body.

Autopsy.—All the large venous trunks enormously dilated. The tissues were apparently " wet."

LXII

Genito-urinary experiment. Manipulation of the ovaries, tubes, uterus, bladder, and kidneys was attended by no change in blood-pressure. Splanchnic shock was induced, then saline solution was infused, causing an immediate rise in pressure, which soon fell to about the same level as before. In this case the records are imperfect, as there was a delay in recording the notes.

LXIII

January 18, 1897.—Male dog. Temperature at beginning, 39° C.; at end, 38.5° C. *First.* In performing the technique of squeezing a tube out of the larynx, according to a method suggested in the removal of intubation tubes, complete temporary inhibition of both respiration and heart-beats followed. Soon the inhibition gradually passed off and the functions were resumed. *Second.* Ligation of the pylorus and a Davidson's syringe attached to the œsophagus. First air, then water, was forced into the stomach, causing its gradual complete dilatation. As the dilatation increased, the heart excursions became longer and less frequent. Finally they became extremely long; then, on relieving the tension in the stomach, the normal stroke reappeared. *Third.* Clamping the aorta just below the arch was followed by sudden rise in blood-pressure, without compensatory fall. *Fourth.* Extensive and severe manipulations of the intestines caused very slight rise in blood-pressure. Respirations were extremely altered, as before; then (*Fifth*) mutilating the hind extremities was followed by a slight rise in the pressure. During this extensive cutting the respirations became slower and slower, and finally the dog, in a general muscular contraction, ceased breathing. The heart-beats at this time were vigorous and executing long strokes. The heart continued for some time thereafter.

LXIV

Male dog. *First.* Experiment on the stomach, as in LXIII, repeated. In this case the stomach was allowed to escape from the abdominal cavity.

There was a gradual increase in the length of the heart contractions and the cardiac pause. Finally the stomach burst, and at once the normal excursions were resumed. *Second.* Clamping the aorta, as in LXIII, similar results were noticed. Extensive mutilations and repeated ampu-

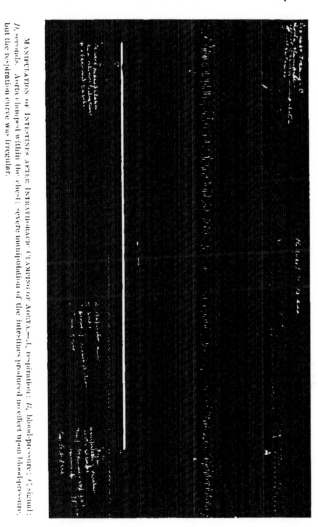

MANIPULATION OF INTESTINES AFTER INTRATHORACIC CLAMPING OF AORTA.—*A*, respiration; *B*, blood-pressure; *C*, signal; *D*, seconds. Aorta clamped within the chest; severe manipulation of the intestines produced no effect upon blood-pressure, but the respiration curve was irregular.

tations through the hind extremities caused marked effect upon respirations. They became very slow, and finally ceased, while the heart beat strong and forcibly for some time. Respiration did not resume, not even a spasmodic effort.

MANNER OF DEATH WITH AORTA CLAMPED.—A, respiration; B, central blood-pressure; C, signal; D, time in seconds. Read from right to left. Note the respiratory failure in injury of the extremity. The diminishing curve at right is the concluding tracing of the blood-pressure on the left. The large undulations are due to respiratory influence upon the blood-pressure. Note the slow, full heart-beats at the left and their gradual diminishing in length until complete cessation.

LXV

The experiments on the stomach, uterus, vagina, mucous membrane, ovaries, and tubes, repeated; clamping aorta. All bore out the previous experiment on the same line.

LXVI

Black bitch. *First.* Manipulations of uterus showed marked rise; the pressure on subsequent manipulations mounting up from the point to which it had risen at the end of the preceding one. A fall in blood-pressure was noted while making the abdominal incision. *Second.* While

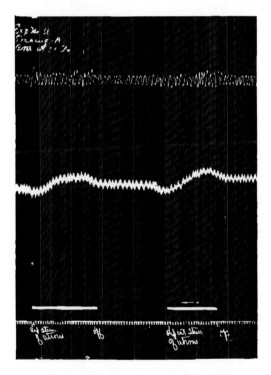

GENITAL ORGANS.—Upper line, blood-pressure; lower, respiration. Uterus manipulated as in performing hysterectomy. Note the rise in the blood-pressure.

clamping the aorta the animal died from failure of respiration. The heart went on beating forcibly and regularly. The animal apparently died twice. The heart and respiration both failed; then, on vigorous artificial respiration, after a lapse of fifty seconds, the heart began beating, and later the respirations were resumed.

LXVII

Large male dog. Experiments on the stomach, testicle, and aorta. Experiment on stomach as in above, with like results. On clamping the aorta the animal soon died of respiratory failure. Heart beat strongly afterwards, as in previous experiment.

LXVIII

January 21, 1897.—Dog; weight, seven and a half kilos; enlarged thyroids. Stomach experiment. *First.* Dilated the stomach by the technique previously described. Light blows upon the partly dilated organ produced a slight fall in the pressure. *Second.* Injected one one-hundredth grain of atropine sulphate, and continued dilating the stomach; there was no change in the character of the pulse-wave during the dilatation. At the moment the stomach finally ruptured there was a sharp fall in the blood-pressure, but no change in the strokes was noted. (See Tracing *B.*) *Third.* Intra-thoracic clamping of the aorta; immediate rise of the pressure, followed by a more gradual further rise. *Fourth.* Injury and manipulation of the intestines and testes produced a slight rise in the blood-pressure. There was a tendency to recurring respiratory failure. Upon continuation of the irritation the respiration became much embarrassed and finally failed. The blood-pressure gradually fell, the strokes becoming shorter and shorter until they ceased at zero.

LXIX

January 22, 1897.—Bitch; weight, nine kilos. Time of experiment, one hour. Temperature at beginning of experiment, 39° C.; at the close, 37.5° C. Stimulating the uterus by traction and pressure, distinct rise in blood-pressure, increased depth in respiration, yet not so marked as in other experiments. (Tracing 69, *A.*) This tracing also shows very good vaginal and bladder curves, indicating rise in blood-pressure. Another tracing (69, *B*) shows the effect of forcing air and water into the stomach to dilate the same. Injection of one one-hundredth grain of atropine sulphate into the jugular vein, then continuing the dilatation, no further changes in blood-pressure or in the character of the heart-strokes were observed. The contents of the stomach were allowed to escape through the œsophagus; slight fall in pressure. Clamping the aorta within the thorax; prompt rise in blood-pressure, followed by a steady fall after the respirations failed. On maintaining artificial respiration there was an increase in the blood-pressure, and after a time the normal respirations were resumed and a high blood-pressure prevailed. After a control began manipulating the intestines; sharp rise followed by some fall, then continued practically unchanged excepting the great respiratory modifications and increased frequency. (Tracing 69, D.) Stimulation of the sciatic nerve produced little if any effect; but repeated stimulations were accompanied

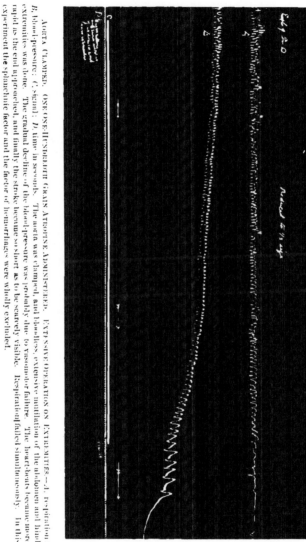

AORTA CLAMPED. ONE ONE-HUNDREDTH GRAIN ATROPINE ADMINISTERED. EXTENSIVE OPERATIONS ON EXTREMITIES.—A, respiration; B, blood-pressure; C, signal; D, time in seconds. The aorta was clamped, and bloodless, extensive mutilation of the abdomen and hind extremities was done. The gradual decline of the blood-pressure was probably due to vasomotor failure. The heart-beats became more rapid as the end approached, and finally the stroke became so short as to be scarcely visible. Respiration failed simultaneously. In this experiment the splanchnic factor and the factor of hemorrhages were wholly excluded.

by respiratory failure, with gradual decline in the blood-pressure and diminution in the length of the stroke until death. Just after stimulation of the sciatic nerve severe injury and mutilation of the hind extremities were made. There was a steady failure of respiration during this time and gradual fall of the blood-pressure as above mentioned.

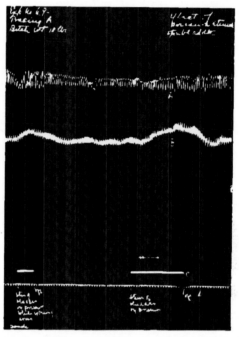

BLADDER EXPERIMENT.—*A*, respiration; *B*, blood-pressure. Note the rise in the blood-pressure and the alteration in the respiratory curve on manipulating the bladder.

LXX

January 23, 1896.—Dog; weight, nine kilos. Time of experiment, one hour and fifteen minutes. Temperature at beginning, 39° C.; at the close of the experiment, 36° C. (Tracing *A*.) *First.* After control, incised the scrotum and exposed the testicle; slight distinct fall in blood-pressure. *Second.* Forcible dilatation of sphincter ani and lower rectum by opening the blades of large scissors; rise in blood-pressure. *Third.* Inserting hæmostatic forceps into nares and making lateral pressure by opening blades produced a rise in blood-pressure; respirations also altered. (Tracing *B*.) *Fourth.* Severe dilatation of nares; rise in blood-pressure. Other results similar to those in Experiment LXIX.

LXXI

January 23, 1897.—Bitch; weight, four and a half kilos. Time of experiment, thirty-five minutes. *First.* Clamped the aorta within the thorax. *Second.* Dilated the nares, stimulated the uterus; no marked effects.

LXXII

January 25, 1897.—Dog; weight, six kilos. Time of experiment, fifty-five minutes. Incised the scrotum and exposed the testes. Lateral traction upon the testes and spermatic cord produced but little effect; slight fall in respiratory curve and a decrease in the amplitude of excursions. Blood-pressure, slight rise, followed by a slight fall, then increase in the amplitude of excursions and a general tendency towards an increase in blood-pressure. Respirations began to fail. Testes were removed by traction, spermatic cord tearing through. Blood-pressure fell slightly, amplitude increased, respirations became less frequent but deeper. Clamped the aorta within the chest, destroyed the stellate ganglion and its communications, and operated extensively upon the hind extremities and posterior portions of the trunk. Respirations began to fail rapidly, blood-pressure failed slowly but steadily. Amplitude of individual excursions but slightly changed. The heart was beating full and strong for some time after respirations had entirely failed.

LXXIII

January 25, 1897.—Bitch; weight, six kilos. Time of experiment, one hour and fifty minutes. Forcible dilatation of the vagina produced a gradual increase in blood-pressure; respirations more shallow, followed by strong expiratory efforts; then irregular respirations followed by another strong expiratory effort, after which the respirations became gradually fuller and deeper. Dilatation of the nares produced slight rise in the blood-pressure; amplitude of respiratory excursions unaffected; rate of respirations nearly twice as frequent. Forcible dilatation of anus and lower rectum produced a higher blood-pressure, which was sustained for some time, and then gradually fell to former level. Respirations became quite irregular, and at the cessation of the dilatation strong expiratory efforts were noted, soon after which the respiratory curve conformed to that before the experiment. Abdominal incision to expose the uterus produced slight fall in blood-pressure, followed by a return to former level. Immediate stimulation of the uterus produced but little effect,—just a slight rise in blood-pressure. A later stimulation produced a more marked rise, which was soon followed by a rather marked decline. Clamping the abdominal aorta within the chest was followed by a rise in blood-pressure. Stellate ganglia were destroyed. On cutting extensively the hind extremities and posterior portion of trunk, the blood-pressure fell slightly for a time, and then for a longer period remained at a level. The heart-beats were full and strong. Respirations during the first part of this period were shallow and irregular. During the second and third parts respirations became deeper and slower. During the fourth the respiratory rate was increased. In the fifth still further increased; then became slower in the sixth, and then gradually failed. Amplitude of the heart-strokes gradually diminished ; beats regular. Blood-pressure fairly constant during this time. After the tenth revolution of the drum respirations almost entirely

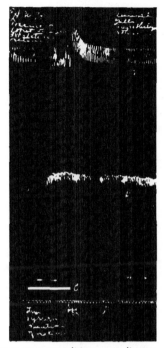

EFFECT OF FORCIBLY DILATING THE RECTUM.—A, respiration ; B, blood-pressure.

INCISING THE SKIN OF THE ABDOMEN. —Upper tracing represents respiration, the middle the central blood-pressure. Note the respiratory change and the decline in the blood-pressure.

ceased ; the heart was still beating regularly, with declining pressure, after respirations had ceased. Releasing the clamp from the aorta, the blood-pressure rapidly fell, and the heart soon ceased beating.

LXXIV

January 26, 1897.—Bitch ; weight, sixteen kilos. The thorax was opened and the stellate ganglion was destroyed. The tracings show the blood-pressure and respiratory action, while there was stimulation by traction and pressure upon the bladder and the ovary ; by stretching the vagina and rectum ; by manipulating the intestines,—very little effect shown by any of the above procedures. Stretching of the anus and the rectum produced slight fall in the blood-pressure and increased depth in respiration. A slight fall was observed upon whipping the intestines.

Electric stimulation of the sciatic nerve produced increased amplitude of heart-strokes, also slight fall in one period of the stimulation.   Made a

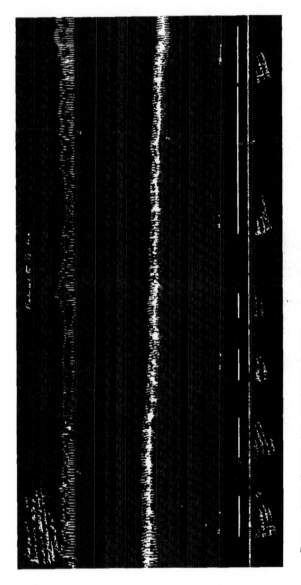

THORACIC AORTA CLAMPED, STELLATE GANGLIA REMOVED.—A, respiration; B, central blood-pressure; C, signal; D, seconds.   In this experiment the aorta was clamped and the stellate ganglion on each side resected before this tracing was made.   Note the even central blood-pressure curve during the infliction of severe injury.

double knee-joint and then double hip-joint amputation.   Respirations gradually failed, followed by a gradual diminution of the amplitude of the heart-strokes, and finally failure of the heart.

LXXV

January 26, 1897.—Dog; weight, seven kilos. Time of experiment, one hour and ten minutes. Incised scrotum; slight fall in blood-pressure. Traction upon testes and spermatic cord; no effect. Opened the thorax; digital pressure upon the base and upon the apex of the heart. Digital

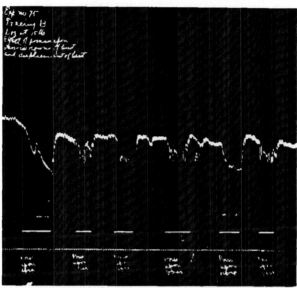

EFFECT OF CONTACT WITH THE HEART.—The tracing in the centre of the cut represents the central blood-pressure. The striking irregularities are due to contact with different portions of the heart as indicated on the cut. Note the sudden drop in blood-pressure, followed by a rapid rise on cessation. Note also the longer strokes in contact with the base as compared with contact with the apex.

pressure upon the base produced long, sweeping excursions, the lever going above and below the line of the control. Pressure upon the apex produced diminution in the amplitude and fall in the blood-pressure. Respirations were diminished and very irregular.

LXXVI

January 27, 1897.—Dog; weight, nine kilos. Time of experiment, one hour and fifteen minutes. Ligated œsophagus at the cardiac orifice. Secured the tube of a Davidson syringe in the upper portion of the œsophagus. Forced in air, thereby rapidly dilating the œsophagus only. There was a sharp rise in blood-pressure; respirations became irregular and very shallow. Upon a second attempt to dilate the œsophagus water was forced through the incompetent ligature, and filled the stomach and a portion of the intestinal tube. During this time there was a gradual rise in blood-pressure. When dilatation became so great as to interfere with respiration the stomach was punctured, which was attended by a sharp fall in blood-pressure, followed by a sharp rise of the same, and then by a gradual decline. After an interval of two minutes, applied a large gas flame to the posterior extremities in the region of the knee; greater excursions of the cardiac lever and gradual decline of blood-pressure after the temporary rise. Repeated the application for a longer time; slight rise maintained for a short time, and then decline to a former level; no change in the character of the heart-beats; no change in the respiratory tracings. Application of a Bunsen flame to the nose produced slight rise in blood-pressure, producing a long, rounded curve.

LXXVII

January 27, 1897.—Dog; weight, seven and a half kilos. Time of experiment, one hour and five minutes. Dilated œsophagus by forcing air

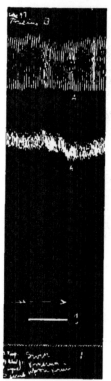

GENITO-URINARY.—*A*, respiration; *B*, blood-pressure; showing effect of pressure upon penis. Note the fall in blood-pressure and the diminished length of the respiratory excursion.

into same after ligature at cardiac end of stomach, as in Experiment LXXVI. Œsophagus was dilated; decided fall in blood-pressure, followed by rise to former level upon cessation of the pressure; the amplitude of respiratory excursions increased. Repeating the same produced a fall of the same character. Upon withdrawal of the tube of the syringe and allowing the œsophagus to collapse there was a sharp and sudden rise in the blood-pressure, followed by a further more gradual ascent. Manipulation of the testes produced a fall in blood-pressure. Repetition of the manipulation and application of great pressure produced a more marked fall. The amplitude of respirations quickly diminished, then gradually assumed their former character. Crushing injury to the penis produced a very similar effect on the blood-pressure and respiration. After this, air was forced into the intestines, producing a gradual rise in blood-pressure; the respirations were unchanged. Manipulation of the distended intestine produced no effect. Repeated forcing

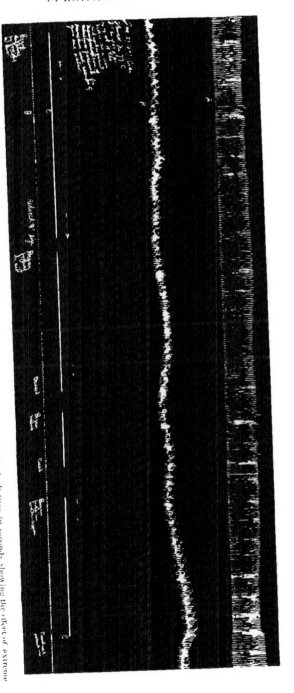

FORCIBLE DILATATION WITH MANIPULATIONS OF INTESTINES.—*A*, respiration; *B*, central blood-pressure; *C*, signal; *D*, time in seconds, showing the effect of extreme dilatation of the intestines by forcing air and water into them. Manipulating the intestines was attended by a gradual rise in the central blood-pressure. The general effect of extreme distension is a gradual rise, as is shown on the right side of the cut. The intestines remained pale, showing that the intra-intestinal pressure was sufficient to prevent dilatation of the splanchnic vessels supplied to the areas so distended, eliminating thereby the splanchnic factor. The rise is probably due to the pressor and cardio-accelerator impulses.

in of water produced a further rise. Puncture of intestines; fall in blood-pressure, increase in cardiac excursions, slight increase in the amplitude of the respiratory movements. Repeated applications of Bunsen flame, no appreciable effect.

LXXVIII

January 29, 1897.—Dog; weight, six kilos. Time of experiment, one hour and a half. Ligated the œsophagus at its cardiac orifice and secured a tube in its upper end. Sudden dilatation of the œsophagus with air produced a sharp fall in blood-pressure and a decrease in the frequency of the heart-beats; at the same time a diminution in the amplitude of the heart-strokes. The respiratory curve showed a marked diminution in depth and frequency. Blood-pressure began to rise before the cessation of the stimulus.

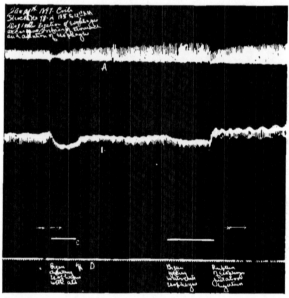

FORCIBLE DILATATION OF THE ŒSOPHAGUS.—*A,* Respiration; *B,* central blood-pressure; *D,* seconds. Note the fall in blood-pressure on dilating the œsophagus, also the slowing of respiration and lessening the amplitude of excursion.

At its cessation there was a sharp ascent to the normal. Respirations increased in frequency and depth before cessation of stimulation. There was a gradual increase of the depth of respiration until the rupture of the œsophagus by forcing in water. During the forcing in of this water there was a slight decline in blood-pressure. At the instant of the rupture there was a sharp, sudden rise in blood-pressure, and thereafter an increase in the amplitude of the beat until nearly twice the former amplitude was attained. After this, dilated the intestines and stomach by forcing air and water into the rectum; a gradual rise in blood-pressure during the dilatation with air and a marked fall upon its cessation. Then began forcing in water; a gradual rise in blood-pressure; later, a more marked ascent, as the intra-intestinal pressure became greater. After a period of forty-five seconds, during which blood-pressure remained unchanged, puncture of the stomach with a scalpel was made; a sudden but slight fall in blood-pressure ensued, with increase in amplitude of the heart-beats, and a slightly increased diminution of their frequency. The mechanical error in the respiratory tracing, an error due to the abdominal expansion, invalidates the respiratory tracings. Injury to the testicle and dissection of the lower extremity to expose the sciatic nerve caused no appreciable effect. Electrical stimulation of the sciatic nerve produced a marked increase in the force of the heart-beats and a moderate rise in the blood-pressure; the depth and the frequency of the respiration were also increased. Mechanical stimulation, especially scraping with a scalpel, in particular the sciatic and median nerves, produced in each instance a marked increase in depth of respiration and ascent in the blood-pressure curve. The heart continued beating for a long time moderately full and strong, while respiration gradually failed and ceased first.

LXXIX

January 29, 1897.—Dog; weight, six kilos. Time of experiment, twenty-five minutes. Incision in the lumbar region to expose kidney; manipulation of the kidney; sharp fall in blood-pressure; the pressure began to rise before manipulation ceased, and so reached its former level. Repetition of the manipulation produced a less marked fall than before. There was a gradual decline for some minutes afterwards.

After a few moments' interval the dog was shot with a .38-caliber Smith & Wesson revolver. The ball passed through the ninth rib, shattering it; through the diaphragm, two lobes of the liver, ascending colon, seven times through the small intestines, and made its exit from the left loin. At the point of entrance into the liver a stellate lesion was noticed, very similar to such stellate fracture of skull. At the time of shooting the blood-pressure slightly rose, followed by a slight fall, the heart-beats gradually increasing in frequency for a period of thirty seconds, when they became slowed.

Respirations.—A strong expiratory effort, followed by shallow respirations, which gradually became deeper and less frequent. A second shot was now fired into the dog, the effect of which was to produce an immediate cessation of respiration. After about thirty seconds a few spasmodic respiratory attempts were noted. After respirations failed the respiratory curve, which showed no change at the instant of shooting, displayed long sweeping excursions, intermingled with smaller strokes at irregular periods. The heart-beats gradually became smaller and failed.

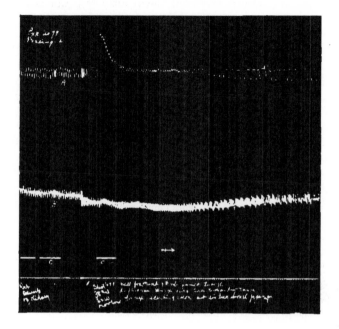

GUSSHOT WOUND OF THE CHEST.—*A*, respiration; *B*, central; *C*, signal; *D*, seconds. At the third signal-mark the animal was shot. The ball fractured the ninth rib, passed through the diaphragm, through the liver, twice through the duodenum, through the ascending colon, and five times through the jejunum.

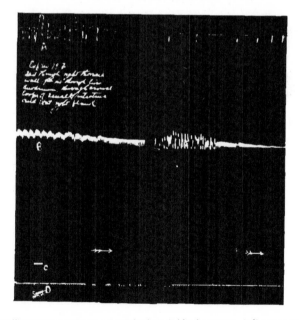

GUSSHOT WOUND OF THE CHEST.—*A*, respiration; *B*, blood-pressure. Ball passed through right thoracic wall, ninth rib, liver, and duodenum, several loops of small intestine, and made its exit through right flank. Note respiratory effect.

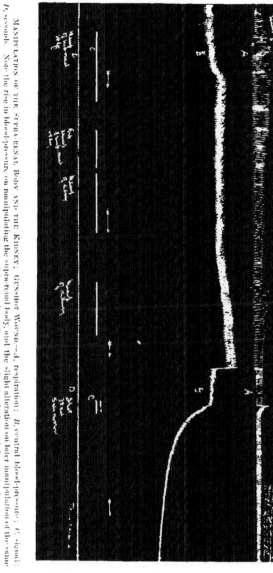

LXXX

January 29, 1892.—Bitch. Time of experiment, one hour and fifteen minutes. Manipulation of the supra-renal body produced a sharp ascent in the blood-pressure. Repeated manipulation of this body and manipu-

lation of the kidney produced no appreciable effect. After securing a
control, the dog was shot with a .38-caliber Smith & Wesson revolver.
The ball passed through the tenth rib, the liver, six times through the
small intestines, through the rectum, and finally passed out through the

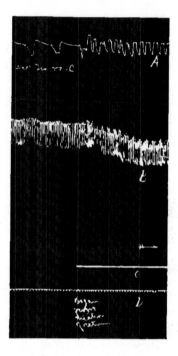

DILATATION OF RECTUM.—*A*, respiratory curve; *B*, blood-pressure curve; showing effect on
a failing respiration of stretching sphincter ani and lower rectum. The effect upon the blood-
pressure was but temporary. Read from left to right.

left flank. Blood-pressure; a sudden sharp decline, followed by a more
gradual descent of diminishing rapidity, producing a full rounded curve.
The amplitude of the beat gradually diminished as the blood-pressure fell,
then became quite irregular, and failed coincidently with the respiration.

Respiration.—Depth and frequency greatly diminished after a short
period of acceleration following the strong expiratory effort which occurred
at the instant of the shot.

LXXXI

January 30, 1897.—Dog; weight, eight kilos. Time of experiment,
one hour and a half. Exposed kidneys by lumbar incision. Manipulation of supra-renal body produced a sharp rise in blood-pressure, followed
by a quick return to its former level. Respirations were slightly irregular
and increased in frequency. Slight rise in blood-pressure upon manipu-

lation of the kidneys. Stripping of capsule and traction produced little effect. Compression of renal vessels produced slight fall. Exposed the testicle and injected atropine. Severe pressure upon the testicle produced but little effect on the blood-pressure,—a slight fall. Respirations became

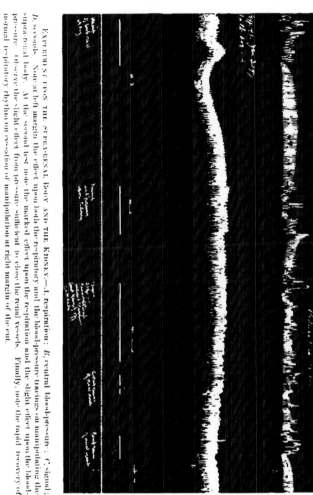

EXPERIMENT UPON THE SUPRA-RENAL BODY AND THE KIDNEY.—A, respiration; B, central blood-pressure; C, signal; D, seconds. Note at left margin the effect upon both the respiratory and the blood-pressure tracings on manipulating the supra-renal body. At the second test note the marked effect upon the respiration and the slight effect upon the blood-pressure. Observe the slight effect from pressure sufficient to close the renal vessels. Finally, note the rapid recovery of normal respiratory rhythm on cessation of manipulation at right margin of the cut.

more shallow; frequency unchanged. Shortly after this, respirations became quite shallow and slow; blood-pressure strokes were much increased in amplitude. Dilatation of the anus and the rectum produced an increased frequency and amplitude of the respiratory movements. Blood-pressure fell gradually at the beginning of the dilatation. The rate of the heart-beats increased ; then they returned to the former rate with diminished force, but soon recovered their original rate and force of beat. Repetition of dilatation of the rectum produced very little effect upon the blood-pressure. Respiration showed an immediate change,—increase in depth and frequency. Application of Bunsen's flame produced a rise in the blood-pressure and increased force of heart-beats; the depth and frequency of respiration also increased. In this experiment there was considerable trouble with clotting in the carotid canula. The canula was cleared no less than ten times.

LXXXII

January 30, 1897.—Bitch ; weight, four kilos. Time of experiment, one hour and fifty-five minutes. One-third of a grain of curare and one-twelfth of a grain of morphine were injected into jugular vein. This was followed by an immediate fall in blood-pressure and a gradual decrease in the force of the heart-beats. Soon the cardiac activity became almost imperceptible, the respirations had failed, and the animal was thought to be dying. Artificial respiration was begun. After a considerable interval, during which artificial respiration was maintained, there was a marked increase in the force of the heart-beats, the rate somewhat slow. Later a slight increase in the rate, with a marked increase in the amplitude of the stroke and a considerable recovery of the blood-pressure. Upon manipulation of the supra-renal body and kidney a slight rise in the blood-pressure ensued, followed by a prompt return to its former level upon cessation of the stimulus. On repetition, the same stimuli produced very little, if any, effect. Application of Bunsen's flame to the right hind foot caused an appreciable rise in the blood-pressure. After this there was a gradual diminution in the force of the heart-beats, until the carotid tracing became very small. On stopping artificial respiration the heart-beats became less frequent but somewhat stronger, and then gradually declined and stopped. Artificial respirations were maintained during two revolutions of the drum.

LXXXIII

January 31, 1897.—Bitch ; weight, five kilos. Secured control curve, then shot the animal with a .38-caliber Smith & Wesson revolver. The ball passed into the body at the level of the ninth rib of the right side, passed through the diaphragm, through the vena cava inferior, the lower duodenum, five times through the jejunum, and made its exit through the left flank. At the instant of the discharge of the revolver blood-pressure fell rapidly ; rapidity of descent gradually decreased and soon its fall ceased. Pulse showed some decrease of force at first, then an increase in its force and a considerable decrease of irregularity in both force and frequency. Then the blood-pressure gradually declined, the rate increased and the force diminished. The respiratory and the circulatory action failed simultaneously.

LXXXIV

January 31, 1897.—Dog; weight, five kilos. Shot with a .32-caliber Smith & Wesson revolver. Ball passed in at the level of the sixth rib on the left side, fracturing the same and driving a spiculum of bone into the pleural cavity and into the lung, thence passed through the lower border of the lung, the diaphragm, the cardiac end of the stomach, touched there the cardiac orifice, thence through the diaphragm into the pleural cavity. In its further course it cut through the greater splanchnic nerve, thence through the aorta, entering the left, and escaping through the right wall of this tube. At the orifice of its exit it produced a peculiar stellate hole, the intima showing this condition especially well. Its further course was through the right side

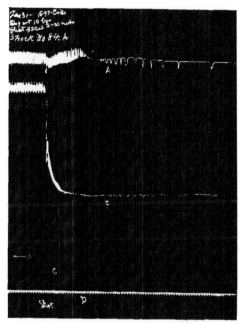

GUNSHOT INJURY.—Death from venous hemorrhage after shot injury. See Protocol LXXXIV. *A*, respiration; *B*, blood-pressure; *D*, seconds.

of the diaphragm, through one of the lobes of the liver, and thence through the right kidney and out through the right loin. At the instant the shot was fired there was a very great and sudden fall in the blood-pressure. The heart failed rapidly and soon ceased. At first respirations were much accelerated, but soon they became shallow and infrequent and quickly failed—failed before the heart did.

LXXXV

February 1, 1897.—Dog; weight, twenty kilos; large black animal, very powerful. Encountered considerable difficulty in anæsthetizing him. Blood-pressure was two hundred and ten millimetres at beginning of experiment. Began the operation for the purpose of destroying the accelerating fibres to the heart. While making an opening into the thorax noted that the blood was quite dark. Artificial respirations were at the time being maintained. The anæsthetic was withdrawn and the heart rapidly failed. The dog died of cardiac failure during artificial respiration.

LXXXVI

February 2, 1897.—Dog; weight, seventeen and a half kilos; large, black animal, very much like the one used in Experiment LXXXV. While performing same operation as in LXXXV, and at the same point in the experiment, the dog died in a similar manner. In this case the blood did not appear dark.

LXXXVII

February 2, 1897.—Dog; weight, seventeen and a half kilos. Time of experiment, two hours and forty-five minutes. Injected curare and morphine into the jugular vein; artificial respiration maintained. Severed the cardiac branches of the stellate ganglion; first severed the right, then the left. Heart's action quite irregular while cutting the right fibres. While cutting the left fibres the heart's action was quite weak and blood-pressure fell. A sharp rise after the fibres were cut was noted, followed very soon by a fall and by increase in amplitude of the strokes. Temporary cessation of respiration was attended by a rise in blood-pressure and decrease in the amplitude of the heart-strokes. Any little irregularity of the artificial respiratory rhythm produced quite a marked effect on the blood-pressure, and thereby greatly obscured the results of the experiments. The sciatic nerve was exposed and stimulated by a faradic current. A sharp increase in blood-pressure during the period of stimulation was noted. The operation for severing the cardiac branches of the stellate was quite tedious, and the animal lost considerable blood from continuous oozing. Hemorrhage seemed to be excessive wherever the tissues were cut. Very large and vascular thyroid bodies were noted.

LXXXVIII

February 3, 1897.—Dog; weight, nine kilos. Time of experiment, thirty-five minutes. Connected the arterial and respiratory apparatus in the usual manner, and an additional mercury manometer to record the tracing of the superior mesenteric vein, as the first experiment in determining the changes of pressure in the splanchnic area. Manipulation of supra-renal body produced very little, if any, effect. Manipulation and pressure upon the kidneys, incising and crushing the kidneys, no effect,—possibly a slight rise mechanically (?). No effect ensued upon manipulation of the other supra-renal body. Severe manipulation of the other kidney produced slight rise in blood-pressure and some irregularity in respiration. Slight rise in blood-pressure during severe manipulation of the eye; respiration unchanged. At this point respirations began to fail —became shallow and irregular. Severe manipulation of the tongue, puncture, crushing, etc., produced no effect on the blood-pressure. Puncture of the tongue was followed by a few deep irregular respirations. Artificial respiration was supplied at intervals, when breathing was bad. Faradic stimulation of the sciatic nerve produced a slight increase of the blood-pressure and a fuller, slower pulse. The mercury manometer was not sensitive enough for recording the portal pressure.

LXXXIX

February 3, 1897.—Dog; weight, nine kilos. Manipulation of suprarenal bodies, slight rise in the blood-pressure, respiration more shallow and irregular. Manipulation of the kidneys, no effect upon blood-pressure or respiration. Manipulation of liver. There was a sharp fall just before the manipulation of this organ and following the severe injury to the kidney. Manipulation of the spleen produced a slight rise of the blood-pressure and irregular respiratory movements. The dog was now shot with a .32-caliber Smith & Wesson revolver, the ball passing in at the level of the third rib, fracturing the same two inches from the mediosternal line on the right side, through the upper lobe of the right lung, out of the fifth intercostal space of the right side. The drum was not moving when the shot was fired, so the immediate effect on cardiac and respiratory action was lost. Mechanism was started immediately: respiration full, deep, and of normal frequency; heart-beats scarcely perceptible. Heart's action soon became more marked, but was quite irregular, blood-pressure falling meantime. After an interval a second shot was fired, which passed through the third intercostal space, close to the right border of the sternum, through the trachea, œsophagus, and thoracic duct, and passed out of the fifth intercostal space, just to the left of the vertebral column. Postmortem showed that the stomach was greatly distended with air. It was thought to have found its way in through the communication which the bullet established between the trachea and the œsophagus. After this second shot the blood-pressure continued to fall, the heart's action became more feeble, respiration more infrequent, and stopped before the heart did.

XC

February 4, 1897.—Dog; weight, five kilos. Time of experiment, one hour and twenty minutes. Began incision to open the abdomen. Noted an immediate fall in the blood-pressure as the skin was incised. While cutting through the abdominal muscles the blood-pressure continued at a uniform level, and was unchanged by the procedure. Upon cutting through the parietal peritoneum a marked fall in blood-pressure occurred ; respirations became more shallow following the fall. Began incising the peritoneum ; blood-pressure began to rise gradually, and continued to do so for some time. Compression of the liver produced a more marked upward tendency of the blood-pressure. Compression of the spleen pro-

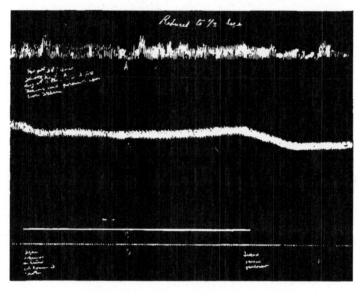

CUTTING THROUGH THE ABDOMINAL WALL AND OPENING THE PERITONEAL CAVITY.—A, respiration: B, central blood-pressure; C, cutting through abdominal wall. Note fall in cutting the skin and on opening abdominal cavity, also respiratory alterations. The peritoneum was incised at the end of the signal, and during the time marked by the signal the abdominal wall was incised.

duced slight rise in the blood-pressure, with a fall to a point below its previous level immediately upon cessation of the compression. Respirations were more shallow during this period. Respirations soon failed ; heart-beats continued strong and regular, but gradually became more and more infrequent until they ceased.

XCI

February 4, 1897.—Dog ; weight, seven and a half kilos. Time of experiment, one hour and forty-five minutes. Injected two-thirds of a grain of curare and one-sixth of a grain of morphine. At the time of injection there was a slight rise in blood-pressure, and in fifty seconds later a small, sharp rise, with a decrease in the frequency and length of the stroke. Artificial respiration was maintained. Began operation upon the thorax with a view to destroying the cardiac branches of the stellate. On opening the thorax the heart became very weak and soon failed. Artificial respiration was of no avail.

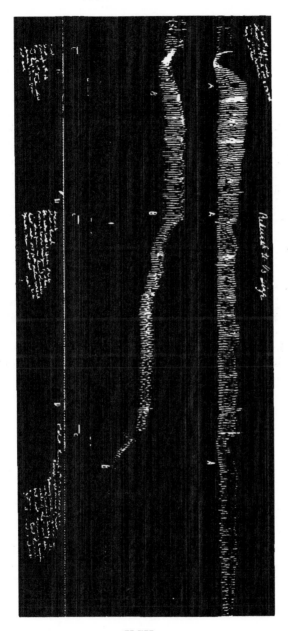

XCII

4, 1897.—Dog ; weight, ten kilos. Time of experiment,
minutes. Shot three times with a .32-caliber revolver. The

first shot passed through the fourth rib, right side, through the right lung, upper lobe, and out through the right sixth intercostal space. Blood-pressure fell rapidly, but the fall soon ceased and the pressure began to rise again, the beats becoming small and rapid. Soon the heart's action became quite normal again. Respiration very shallow, soon returned to normal. Second shot passed through eighth intercostal space on the right side, through the lower margin of the right lung, through the diaphragm and liver, cut the large lymphatic vessels, and passed out through the tenth intercostal space of the right side near the vertebral column. Blood-pressure fell, not so sharply as after the first shot, and the beats were but little changed, their amplitude perhaps slightly decreased. Blood-pressure soon reached its point of greatest decline, and then ran along nearly on the level until shot three was fired. Rate of pulsation increased greatly, beginning about fifty-six seconds after shot two. Respiration practically unchanged by shot two, slightly decreased in depth. Shot three passed through the fifth right intercostal space near the sternum, thence through the pericardium and through the right border of the heart, opening into the cavity of the right ventricle, cutting deeply into the wall of the left ventricle, but not opening the cavity of the same. There was an instantaneous fall of about twenty millimetres in blood-pressure. From this point it ascended sharply for a period of four seconds, and then fell sharply for five seconds, and thereafter declined about one minute, the force of the beats decreasing. Two rapid respiratory movements occurred at the instant of the shot, then a short period of diminished amplitude, after which a return to the former character. Very soon respiration began to fall and soon ceased. The heart manifested feeble activity after cessation of respiration, but it soon stopped.

XCIII

February 5, 1897.—Dog; weight, eight kilos. Duration of experiment, one hour and thirty minutes. The tracing was taken while cutting through the abdominal wall. There was a slight fall in the blood-pressure on incising the skin, also while incising the peritoneum. Respiration became irregular when the peritoneum was opened. Upon the compression of the liver a very slight fall in blood-pressure was noted. Respiration remained unchanged. Severe pinching of the gall-bladder was accompanied by marked fall in the blood-pressure and an irregularity in the respiratory action. While this traction was made, an increase in the volume of the venous trunks nearest to the point of traction was noted. Upon cessation of traction, the retarded blood flowed into the heart and there was an increase in the blood-pressure in the skin. The gall-bladder was manipulated roughly, producing altered respiratory rhythm and immediate fall in blood-pressure. Dilation of the cystic duct with forceps produced no effect. The animal died of hemorrhage from accidental

OPERATION ON GALL-BLADDER.—*A*, respiration; *B*, blood-pressure; *C*, signal; *D*, seconds. Note the irregularity of the respiratory tracing during the manipulation. The immediate fall in blood-pressure was probably due to mechanical interference with the larger venous trunks.

rupture of a large branch of the portal vein. Cardiac and respiratory action ceased simultaneously.

XCIV

February 6, 1897.—Dog; weight, eleven kilos. *First.* Introducing the canula into the femoral vein by the ordinary technique; very considerable decline in blood-pressure; respiration deep and full. Seven hundred cubic centimetres of normal saline solution at 40° C. produced a decided rise in blood-pressure, with increase in the amplitude of the heart-strokes, until two hundred cubic centimetres had been given, after which there was little alteration. *Second.* Electrical stimulation of the sciatic nerve produced marked increase in blood-pressure and increase in the force of the individual heart-beats. Peripheral venous pressure paralleled the central. Respirations were irregular, some full and deep, some shallow and rapid. Application of Bunsen's flame to the foot; rise in blood-pressure. After this, stimulation of the sciatic produced a slight rise. Respirations less deep during the period of stimulation. Repeated application of Bunsen's flame for a period of two minutes produced decided rise in blood-pressure, the peripheral paralleling the central. In a little less than one minute after

the application of the stimulus the blood-pressure reached its highest point, and then commenced a gradual decline, with a more abrupt fall upon cessation of the stimulus. *Third.* The animal shot with .32-caliber Smith & Wesson revolver, ball passing just internal to, and forward of the angle of the jaw, right side, ranged slightly towards the median line, entered the skull just anterior to, and near the exit of the third division of the fifth nerve (foramen ovale); thence passed through the region of the cortex cerebri, which, as a result of previous experiments on localization, we had designated the " respiratory arrest area." The ball then passed upward, forward, and outward, crossing through the hippocampus major and the descending horn of the right ventricle, and escaped by perforating the cranial vault in the median line, at a point one and a half inches posterior to a line

GUNSHOT INJURY.—*A*, respiration; *B*, central blood-pressure; *C*, signal; *D*, time in seconds. Shot passed through the head. Observe respiratory arrest and continuation of cardiac action.

drawn between, and connecting the highest portion of the superior margin of the orbital cavities. The superior longitudinal sinus was cut through, and there was an extensive clot anterior and posterior to the point of exit. Blood tracing shows that the heart made strong irregular movements at the moment of impact of the shot, and then returned to its former character of beat. The blood-pressure fell quite sharply for a period of eight seconds, and then declined more gradually. Venous pressure partook of the same general movement. Respirations stopped absolutely at the instant of the shot, and no attempt at respiratory movement was afterwards made. The shot was fired at short range.

XCV

February 7, 1897.—Bitch; weight, ten kilos. Time of experiment, one hour and fifty minutes. Gave intravenous injection of four hundred cubic centimetres of normal saline solution at a temperature of 42° C.; marked rise in blood-pressure and increase in force of heart-beat. Parallel venous pressures, also nearly parallel to the arterial curve. Forcible dilatation of the vagina produced fall in central pressure and in the peripheral venous. There was some irregularity of the heart's action noted. Respirations shallow, but quickly returning to their former character. Application of Bunsen's flame to the paw produced but slight rise in both pressures, the venous perhaps more plainly marked. Infusion of three hundred cubic centimetres of same solution produced gradual rise of pressure, with increase in amplitude of strokes in the central. This increase continued throughout the entire time of its administration, peripheral venous pressure paralleling the central. Bunsen's flame was now applied to the foot, causing rise in both pressures, with a curve comparable to that noted in Experiment XCIV. Blood-pressure began to fall before the stimulus was discontinued. After the above procedure, during the time occupied in changing drums, blood-pressure had fallen considerably. An additional two hundred cubic centimetres of saline solution caused again a rise in every way comparable to the preceding. Application of Bunsen's flame now produced a sharp rise in the pressures, with strong, full beats and diminished frequency; respirations shallow and irregular. After this the blood-pressure fell. Application of Bunsen's flame to the intestines produced no effect, except a slight change in the character of the heart-beats; also perhaps slightly accelerated a gradual fall in the pressure, which was taking place at the time. Application of the flame to the foot at this time seemed to check the fall in the central pressure. Peripheral pressure continued its downward course unchanged. Injection of two hundred and fifty cubic centimetres of normal saline solution caused a moderate rise in central pressure; peripheral unchanged; respirations unaltered.

XCVI

February 8, 1897.—Dog ; weight, ten kilos. Preliminary preparations as in preceding experiment. The animal was in good condition, the pressures all writing well, but, while adjusting the tubing and about to begin the abdominal incision, the dog suddenly died. (Solution from the pressure-bottle into the circulation.)

XCVII

February 8, 1897.—Dog ; weight, ten kilos. A lank, shaggy-haired, young dog ; was not a good subject. Respirations failed soon after being placed on the board and the apparatus connected. Artificial respiration was necessary during the entire experiment. Soon after commencing the tracing, began to infuse saline solution. Blood-pressure was falling at this time. It continued to fall for two seconds, when artificial respiration was temporarily discontinued because the animal manifested a tendency to spontaneous breathing. Respiratory efforts were practically absent for a few seconds, during which time the blood-pressure rose rapidly. Infrequent and insufficient respiratory action was now noticed. The blood-pressure fell during, and immediately after the respiratory effort. There now appeared a series of the characteristic oscillations in blood-pressure and variations in the character of the heart-strokes. During the decline in blood-pressure the heart-strokes were long and relatively infrequent. During the ascent, the strokes were more rapid and relatively short. Upon resuming artificial respiration the heart-strokes became shorter, more rapid, and the blood-pressure fell rapidly. It, however, soon reached the level occupied at the beginning of the experiment, and then ceased falling. After the lapse of a few seconds, during which time artificial respiration had been maintained and the blood-pressure had remained unchanged, artificial respiration was again discontinued and gave rise to phenomena similar to those above described. Five hundred cubic centimetres of saline solution were injected. The effect of the first one hundred and fifty cubic centimetres was more marked. Application of a Bunsen's flame produced the characteristic increase in blood-pressure. In each instance a decline began before discontinuing the flame. In each instance also the amplitude of the heart-strokes was diminished. During the subsequent part of the experiment one hundred and twenty-five cubic centimetres, and one hundred and seventy-five cubic centimetres of normal saline were injected, causing in each instance a prompt rise in blood-pressure. Stimulation of the sciatic by the faradic current produced an increase in blood-pressure and greatly diminished the amplitude of the heart-strokes. Repetition of the stimulus produced a further rise in blood-pressure, with diminution in amplitude as before. Heart soon failed.

XCVIII

February 9, 1897.—Dog ; weight, nine kilos. Duration of experiment, two hours and eight minutes. Thyroid glands enlarged, and in the dissections the animal bled freely everywhere. The tissues were tough, making it quite difficult to tear or cut the fasciæ. In opening the abdomen, both the skin and the peritoneal incisions produced quite a marked fall in blood-pressure. In attempting to introduce the canula into the portal vein this structure was severely torn, which occasioned some hemorrhage, necessitating clamping the vein. Blood-pressure fell considerably, and the heart-strokes became relatively small. Attempted to infuse saline solution, but found difficulty, and so inserted the canula in another vein and infused two hundred and fifty cubic centimetres at 43° C. temperature, which produced a prompt rise in blood-pressure and a stronger pulse. Application of Bunsen's flame to the paw produced an abrupt increase in the blood-pressure and a diminution in the amplitude of the heart-strokes. Application of Bunsen's flame to the peritoneum produced a fall in the blood-pressure, without any change in the character of the stroke. First infusion of two hundred and fifty cubic centimetres of saline solution produced the usual rise in both arterial and venous pressures. The latter, however, responded but slowly and sluggishly. Infusion of three hundred and twenty-two cubic centimetres produced prompt and marked rise of arterial pressure, with slight and gradual increase in the venous. Respirations were also much increased in amplitude and frequency. Application of Bunsen's flame to the feet was repeated, with the same results as noticed above. Further injections of saline produced a less marked increase in blood-pressure.

In all, one thousand cubic centimetres of solution were injected. At the termination of the experiment the pulse and respiration were fairly good. A small amount of ether was put into the trachea and the tube clamped. After a few attempts at respiration, respiratory movements ceased, the heart continued beating for a period of ninety seconds, then gradually ceased.

XCIX

February 10, 1897. — Bitch; weight, nine kilos. Duration of experiment, two hours and fifteen minutes. Incision of the skin of the abdomen produced a marked fall in the arterial blood-pressure and a moderate rise in the peripheral venous. The arterial blood-pressure soon returned to, and the venous fell slightly below, its former level. Incision of the peritoneum produced but a very slight fall in the arterial pressure. Respiration slightly irregular while making the skin incision. Manipulation of the omentum produced no effect. After some moments the pulse became weak, the blood-pressure declined considerably, and respiration became more

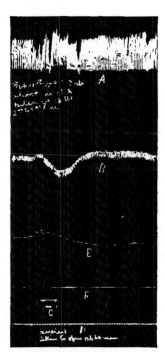

INCISING THE SKIN IN MAKING ABDOMINAL SECTION.—*A*, respiration; *B*, blood-pressure; *E*, peripheral venous in femoral; *F*, peripheral cephalic, arterial. Note the fall in the central after a temporary rise, and rise in the peripheral venous.

shallow and infrequent. Injection of three hundred cubic centimetres of normal saline solution at 43° C. produced a marked rise in blood-pressure, with increased length of heart-stroke. The venous pressure paralleled the central. Dilatation of the vagina produced no effect upon blood-pressure, but respirations were slightly disturbed. Bunsen's flame applied to the right hind foot produced a rise in both arterial and venous pressures. Respirations slightly disturbed, more shallow, and slightly less frequent. Bunsen's flame applied to the peritoneum produced a slight, but a distinct fall in the blood-pressure. Application of the flame to the left forefoot apparently produced the most marked rise in pressure, and also affected respiration to a greater degree. Injection of one hundred cubic centimetres of saline solution produced but a moderate rise in blood-pressure, improving at the same time the character of the pulse. After the lapse of a minute, fifty cubic centimetres of saline solution were introduced, causing a slight rise in blood-pressure. Dissection to expose the sciatic nerve was next begun, during which no alterations in any of the tracings were noted, excepting the respirations were a little more shallow, but fairly regular in rhythm. Electric stimulation of the sciatic nerve produced slight rise in blood-pressure, followed by a prompt fall. Respirations were considerably accelerated, and their depth was not altered. In dying the respirations failed before the heart ceased beating.

C

February 11, 1897.—Dog; weight, six kilos. A lank, shaggy-haired cur. The experiment lasted two hours and fifteen minutes. Incision of the skin of the abdomen produced a decided fall in both the arterial and the peripheral venous pressures. Incision of the peritoneum produced no appreciable effect. At this point the respirations failed and artificial ones were supplied. Fifty cubic centimetres of saline were injected. A gradual rise followed, however, in both the arterial and venous pressures. Exposure of the mesentery and severely manipulating it was attended by a gradual decline in the blood-pressure. Infusions of one hundred cubic centimetres of saline caused a gradual increase of the blood-pressure, with an increase of the amplitude of the heart-stroke. Digital pressure upon the apex of the heart, the pericardium uninjured, produced small, irregular beats and a sharp fall in blood-pressure. Upon cessation of the digital pressure it mounted rapidly to its former level, and the normal character of the beats was resumed. Pressure upon the base of the heart also produced fall in blood-pressure, but the decline was not so great as when the apex was touched, and instead of the small, irregular beat observed when the apex was touched, sweeping irregular strokes during the period of low blood-pressure accompanying the stimulation were noticed. Pinching the lungs near the base of the heart produced irregular, sweeping heart-strokes and a slight decline of the blood-pressure. Picking up the pericardium with a forceps and making slight traction was accompanied by a fall in pressure.

Upon suddenly releasing the pericardium the blood-pressure rose rapidly, and soon reached a point much higher than the former level. It soon declined to the level occupied at the time the pericardium was first picked up. Incision of the pericardium; no effect. A sharp fall in blood-pressure and a short pause in cardiac activity were noted upon puncturing the heart with a scalpel. Infusion of saline solution was made during this time. The blood-pressure gradually fell. Artificial respirations* were discontinued, and the heart soon ceased beating.

CI

February 1?, 1897.—Dog; weight, nine kilos. Appeared to be in good condition, and the experiment was progressing well after the arterial canula and the respiratory apparatuses were connected. In adjusting the apparatus the pinch-cock controlling the pressure-bottle tube was inadvertently released. Infrequent and irregular cardiac action was at once noted.

The magnesium solution flowed from the pressure-bottle, its pressure being greater than the blood-pressure, and immediate failure of the heart was produced.

CII

February 12, 1897.—Dog; weight, nine kilos. Duration of experiment, two hours and sixteen minutes. Exposure and injury of testicle were attended by a sharp fall in arterial pressure, more gradual in the venous. Respirations were irregular and shallow. There was recovery after a lapse of about one minute. The skin and peritoneal incisions in making abdominal section produced but little alterations,—a slight fall in the blood-pressure. Mechanical irritation of the parietal peritoneum caused a slight fall in

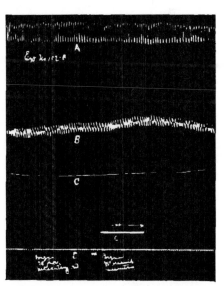

MANIPULATION OF THE OMENTUM.—*A*, respiration; *B*, central blood-pressure; *C*, signal; *D*, time in seconds, showing slight rise in blood-pressure; no effect upon respiration on manipulating the omentum, in contrast with the fall on manipulating the intestines. (An error in the cut—instead of mesentery substitute omentum.)

blood-pressure; upon manipulation of the omentum a slight rise occurred. Practically no change in respiration. Through the abdominal incision loops of the intestine were drawn. At first there was a rise in pressure, which was followed by a gradual fall. Beating the intestines caused an irregular blood-pressure curve. Infusion of seventy-five cubic centimetres of saline produced a gradual and marked increase in pressure. The animal was now shot through the right hind limb with a .32-two caliber Smith & Wesson revolver. After a short interval a slight rise in blood-pressure, followed by a more marked decline, was noted. At the instant of firing the respirations became slightly irregular. Afterwards the animal was shot twice. The first shot passed external to the larynx and penetrated the fleshy tissues of the neck. The second passed through the cricoid cartilage and the œsophagus. Respirations stopped almost immediately after the first shot. At this point the tube connecting the carotid became kinked, destroying the value of the arterial curve. The other shot passed in just internal to the angle of the jaw on the right side, upward and inward and slightly forward, entered the cranial cavity by piercing the base of the skull one-fourth of an inch posterior to the foramen ovale. It then entered the temporo-sphenoidal lobe, passed through that structure and through the posterior portion of the right lateral ventricle, then upward, crossed the median line, and escaped from the cranial cavity through the skull, just to the left of the superior longitudinal sinus. Respirations stopped immediately upon firing the shots. The heart continued beating, and after ten seconds the blood-pressure began to rise sharply, soon began a continuous decline until it stopped, five minutes after respirations had ceased.

CIII

February 13, 1897.—Dog; weight, six kilos. A young animal. Took the anæsthetic badly; respirations failed just after the adjustment of the carotid canula. As the cardiac action was full and strong, a little time was allowed to see whether or not the respirations would be normally resumed. The appearance of vasomotor curves and long, sweeping strokes indicated danger. Artificial respiration was begun. Heart's action entirely ceased at this point for a period of thirty seconds, during which time artificial respiration was maintained. After this quiescence of the heart it made one stroke. After another period of thirty seconds it began infrequent beats, which gradually increased in amplitude. Later the frequency and the pressure increased, and the heart gradually returned to its normal action. On beginning the second tracing the animal manifested a tendency to breathe, and artificial respiration was discontinued. Spontaneous respirations were infrequent and shallow, but the respiratory effect upon the blood-pressure was very marked. After securing the control, the intestines were withdrawn; following this, the arterial and venous pressures began to decline, respirations were irregular. There was a mechanical factor in producing this irregularity, on account of returning the intestines to the cavity. The blood-pressure increased after their return, the amplitude of the strokes diminished, respirations became more frequent and irregular. Injection of three hundred and seventy-five cubic centimetres of saline at 42° C. produced increase of pressure. Unfortunately, a clot in the carotid canula obscured the results at this point. After removing

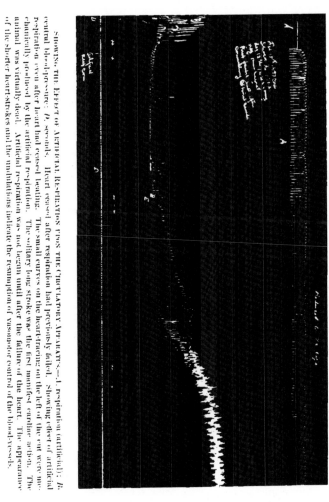

Showing the Effect of Artificial Respiration upon the Circulatory Apparatus.—A, respiration (artificial); B, central blood-pressure; P, seconds. Heart ceased after respiration had previously failed. Showing effect of artificial respiration even after heart had ceased beating. The small curves on the heart-tracing on the left of the cut were mechanically produced by the artificial respiration. The solitary long stroke was the first manifest cardiac action. The animal was virtually dead. Artificial respiration was not begun until after the failure of the heart. The appearance of the shorter heart-strokes and the undulations indicate the resumption of vasomotor control of the blood-vessels.

the clot it was noted that respiration had failed and the blood-pressure had declined. Artificial respiration caused a gradual upward tendency of the blood-pressure line. Temporary cessation of respiration produced no change, gradual upward tendency of blood-pressure continued. After forty-five seconds resumed artificial respirations; then began the infusion of two hundred and seventy-five cubic centimetres of saline solution at 43° C. After one hundred and twenty-five cubic centimetres, artificial respiration was discontinued; blood-pressure soon began to slowly fall. One minute after cessation of artificial respiration voluntary respirations began, and continued with slight irregularity. In five minutes the blood-pressure of the venous began to decline. Digital contact with the apex of the heart produced at first a slight rise, then a sharp and greater fall in blood-pressure, after which the heart's action ceased. Respiration became more shallow and infrequent, the heart remained quiescent for twenty seconds, then made a single stroke. In five seconds, another; in seven seconds, another; then quiescent for thirty seconds; then began to beat irregularly and infrequently. Three hundred and fifty cubic centimetres of saline at 43° C. produced some improvement in cardiac and respiratory action, with slight increase in blood-pressure. Then the respirations became more irregular, heart-action also irregular, after which small beats were resumed. Cardiac and respiratory action ceased simultaneously.

CIV

February 15, 1897.—Bitch; weight, four and one-half kilos. Duration of experiment, one hour and twenty-five minutes. A very scrawny, ill-looking dog, with a tremendous muzzle, out of all proportion to the size of the animal. Soon after putting the animal on the board and connecting the canulæ respirations failed; then the heart ceased beating, and remained quiescent for fifty seconds. Artificial respiration was maintained. Central blood-pressure fell nearly to the venous level. The heart began irregular, infrequent beats; then long, sweeping strokes appeared. The frequency increased and amplitude diminished with a rising blood-pressure. The venous pressure then described a corresponding rise. For a considerable time nothing of importance was noted. During this time the canulæ were being adjusted. Artificial respirations were still maintained.

In the next tracing spontaneous respirations were substituted, showing at first irregular and infrequent strokes, which gradually improved. Infused two hundred and fifty cubic centimetres of saline. At first the arterial pressure rose slightly, then began to decline; the venous pressure was rising while the arterial pressure was falling. A second and a third infusion of one hundred and fifty cubic centimetres each produced the same phenomena, but to a less marked degree. It was thought that the small pulse-beat was due to a clot in the canula, but inspection revealed no clot. Later, respiration began to fail and the blood clotted. The pressure fell, and long, sweeping strokes foretold danger. Artificial respiration was not supplied and the heart's action noted. It soon ceased beating; then artificial respiration was begun, but the heart did not resume its action. After some moments Bunsen's flame was applied to the foot to see whether vasomotor effects might be shown. No perceptible effect noted.

CV

February 15, 1897.—Dog; weight, four and a half kilos. Time of experiment, one hour and ten minutes. A weak, scrawny animal. Respirations failed early and artificial respiration was supplied. A saline injection of two hundred and fifty cubic centimetres at 42° C., producing an increase in the blood-pressure and in the amplitude of the stroke, was given. The animal showed a tendency to natural breathing, and the artificial was discontinued. At intervals voluntary respiration occurred. The intervals between these efforts gradually increased, but respiration was more shallow. The pulse was full and strong. Then began infusion of saline solution, introducing amounts of fifty cubic centimetres at short intervals, until four hundred had been given. During the time of the administration of the last fifty cubic centimetres the amplitude of the heart-strokes became very great. Soon a short pause, then a beat of considerable amplitude, followed by a weak beat. This phenomenon was repeated several times, and then the heart ceased beating, respirations having practically failed some time before.

CVI

February 16, 1897.—Dog; weight, fourteen kilos. Time of experiment, one hour and ten minutes. Respirations failed soon after putting the animal on the board. Heart's action soon became irregular; then for a considerable period the following phenomena were observed. Several irregular sweeping excursions, then for some seconds the heart remained at rest, the only movement traced being that produced by the artificial respirations. Later, the interval between the periods of cardiac activity increased, then slow, sweeping beats at comparatively infrequent periods occurred, but the pressure increased slowly; then for a period the heart's action became more rapid, approaching the normal character. The heart's action soon ceased abruptly, and no beats occurred for a period of eleven seconds, followed by irregular, full, sweeping strokes for twenty seconds, then cessation. After a lapse of eight seconds irregular, full beats began. The heart's action soon improved, and closely approached the normal in character. After another short period of irregularity the heart's action became good, and continued so for a period of two minutes, then the irregularities reappeared during the entire time artificial respiration was maintained at a uniform rate. In every instance a gradual rise of pressure followed diminution or cessation of respiration. Fifty seconds after cessa-

SHOWING THE EFFECT OF ARTIFICIAL RESPIRATION ON BLOOD-PRESSURE.—*A*, respiratory curve; *B*, blood-pressure; *E*, peripheral venous pressure; *D*, seconds. Respiration failed in the course of the experiment. The respiratory curve represents artificial respiration after the gradual cessation on the left. Note the re-establishment of the cardiac action after its failure, by the effect of vigorous artificial respiration; also the rise in pressure and the appearance of longer and more marked vasomotor curves on cessation of respiration. Observe the occasional omission of a beat as indicated by the long stroke.

Showing the Effect of Artificial Respiration upon the Circulatory Apparatus.—A, respiration; B, central pressure; E, peripheral venous in femoral; D, seconds. Showing the effect of artificial respiration. Observe the long period of cardiac activity without respiration. The increasing vasomotor curves and slow sweeping beats foreshadow failure. Observe the effect of resuming artificial respiration after the heart had ceased beating for almost a minute. Artificial respiration throughout where any is indicated.

tion of respiration marked vasomotor curves of increasing length followed. These waves were substituted by long, sweeping strokes, accompanied by a fall in blood-pressure, and finally by a cessation of the heart's action for forty seconds, and then, on the application of vigorous artificial respiration, the beats reappeared. Animal was finally allowed to die of respiratory failure. The dog was a shaggy-haired animal in impaired health, of mixed breed—likely a little spaniel blood.

CVII

February 17, 1897.—Dog; weight, eight kilos. Shaggy-haired cur in impaired health, about two years old. At the one hundredth experiment a bottle reservoir was used in administering the anæsthetic in this and the subjects to one hundred and ten. Respirations ceased soon after placing the animal on the board. For a time there was very rapid respiration, gradually diminishing until its cessation. Meanwhile the blood-pressure had fallen, and the heart's action became weak and irregular. Artificial respiration was begun, but very soon the heart's action ceased and remained quiescent for fully one minute, when infrequent, irregular beats appeared. The force and frequency of the beats were increased and the blood-pressure rose. Saline infusion was begun just after the heart ceased beating, and continued until eight hundred and seventy-five cubic centimetres had been given. Application of Bunsen's flame to the paw caused a rise in pressure in the central, peripheral, and portal, produced slight fall in the central, a more marked decline in the portal, and no change in the peripheral cephalic and peripheral femoral. There was some doubt as to whether a reliable expression of the various pressures was obtained, on account of difficulties in the technique.

CVIII

February 19, 1897.—Dog; weight, six kilos. Blood-pressures were taken in the central carotid, peripheral carotid, peripheral femoral, and central portal, subjecting the animal to an extensive technique before a record was taken. Incising skin over testicle produced a fall in central and peripheral pressures. The portal did not respond properly. Incision of the other side of testicle produced a fall similar to that above mentioned. Severe injury of testicle produced very little effect. Withdrawal of intestines produced but little effect. A fall in the portal pressure while returning the intestines followed, but immediate rise was noted soon after their return. Injected one hundred and fifty cubic centimetres of saline, producing a gradual upward tendency in all the pressures. An additional one hundred and twenty-five cubic centimetres caused an additional rise. Bunsen's flame applied to the foot caused a sharp rise in the central ;

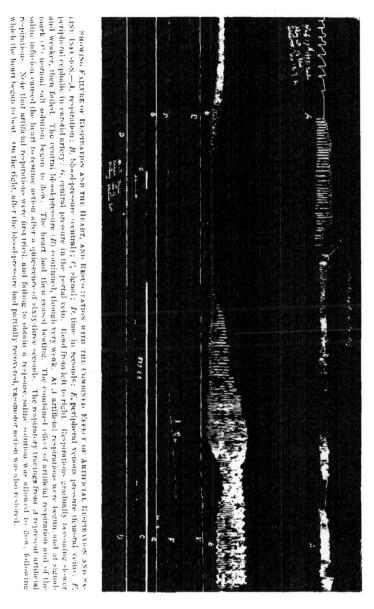

SHOWING FAILURE OF RESPIRATION AND THE HEART, AND RESUSCITATION WITH THE COMBINED EFFECT OF ARTIFICIAL RESPIRATION AND SALINE INFUSION.—A, respiration; B, blood-pressure (central); C, signal; D, time in seconds; E, peripheral venous pressure (femoral vein); F, peripheral cephalic in carotid artery; G, central pressure in the portal vein. Read from left to right. Respirations gradually becoming slower and weaker, then failed. The central blood-pressure (B) continued, though very weak. At A artificial respirations were begun, and at signal mark (C) normal salt solution began to flow. The heart had then ceased beating. The combined effect of artificial respiration and of the saline infusion caused the heart to resume action after a quiescence of sixty-three seconds. The respiratory tracings from A represent artificial respirations. Note that artificial respirations were first tried, and failing to obtain a response, saline solution was allowed to flow, following which the heart began to beat. On the right, after the blood-pressure had partially recovered, vaso-motor action was also restored.

the peripheral and the portal also rose. Dilatation of the rectum produced no appreciable effect. Electric stimulation of the sciatic nerve caused a decided rise in the central and increased the amplitude of the heart's strokes. The portal showed a corresponding change. Stab-wounds in the larynx were made by quick, rapid thrusts ; no appreciable change, excepting that respiration became more irregular and infrequent. The respirations soon failed, and subsequently the heart ceased beating.

CIX

February 22, 1897.—Dog ; weight, five kilos. A poor, scrawny beast, with very large thyroids, large carotid and superior thyroid arteries. In the dissection in this technique there were everywhere free hemorrhages. Respirations failed while finishing connecting the apparatus. The bottle method of anæsthesia was again used. Artificial respiration was maintained and the experiment continued. Four manometers were employed in this case, as before. Intravenous injection of one-thirtieth of a grain of strychnine sulphate was followed, after a period of thirty seconds, by increased force of the heart-beat and rise in the central and portal blood-pressure. The periphery was apparently not working well. After a short period of strong heart-beats there was diminution of the amplitude of heart-stroke but greater rapidity. Then followed a period during which the heart-strokes gradually became stronger and the rate slightly slower. For a considerable period of time, during which the blood-pressure remained at a higher level than before the injection, the pulse gradually became less frequent. Now began infusion of three hundred cubic centimetres of normal saline at 41° C. ; central, peripheral, and portal rose promptly, frequency of heart-beats increased, and amplitude of strokes diminished. After two hundred and fifty cubic centimetres of saline had been introduced, irregularity of the heart appeared ; the rise gave way to a beginning fall. During this entire time there was no disposition towards establishing natural respiration. The heart's action soon became very irregular and ceased.

Post-Mortem.—Quite extreme general venous engorgement. The tissues of the intestines not much filled with blood, but the larger venous trunks were quite full. The venæ cavæ—inferior and superior—were extremely distended, and imparted a sense of very considerable tension to the touch. The heart was in systole, the left auricle and both ventricles were empty, the right auricle contained a small quantity of blood and a small clot. The lungs were comparatively pale and bloodless. The pulmonary vessels contained comparatively less blood than the peripheral vessels. The liver appeared congested and was hard, and upon cutting into it a large amount of bluish, watery blood flowed freely. The kidneys contained far less blood, comparatively, than did the liver.

CX

February 23, 1897.—Bitch; weight, fifteen kilos. At the one hundredth experiment two new factors were introduced : one was the first use of a lot of new ether imported from Germany ; the other, another method of anæsthetizing, which consisted in attaching the rubber tubing of the breathing apparatus to one canula in the cork of a bottle, in the bottom of which a piece of cotton wool received the ether, and another canula in the cork supplied fresh air. These canulæ were of ample size, allowing free exchange of air. After successive respiratory failures which followed hyperpnœa, the respiratory rhythm became gradually slower and slower and more shallow ; finally, with a convulsive twitching of the extremities, the respiration failed, and rarely afterwards showed any disposition to natural breathing. The phenomena in each case were almost identical in this series of ten experiments, and, although great care was exercised, they could not be prevented. At first the ether was suspected, and Squibb's was employed with no better results. In this experiment the bottle was discarded, and the dog anæsthetized through the funnel in the ordinary way with the German ether. The animal stopped breathing ten minutes after being put upon the board. Artificial respirations were supplied but a few moments, when normal respiratory efforts were noted, and natural supplanted artificial respiration. The central peripheral and portal pressures rose slightly. Voluntary respiration was insufficient and artificial was resumed. A second discontinuance displayed similar phenomena, and artificial respiration was again resumed. Practically no changes were noted as the following procedures were carried out in the order named : Incision of the skin of the abdomen.; incision of the muscular walls ; of the peritoneum ; exposure and severe manipulation of the bladder. The animal now began normal respiration, and the artificial was discontinued. Severe dilatation of the vagina was followed by a rise in the central pressure. Some irregularities in the peripheral and the portal made no marked change in their level. Respiration diminished. Severe dilatation of the rectum produced a rise in the central, peripheral, and portal pressures. Respiration was practically *nil* during the time of stimulation. Manipulation of the uterus and the oviducts produced a rise in the central, slight rise in the peripheral and the portal, and the respirations became irregular and shallow. Withdrawal of the intestines was followed at first by a slight fall in the central, then a slight rise, which appeared also in the portal and the cephalic. A slight downward tendency in the peripheral was noted. Soon after these proceedings the pulse became weak ; the central pressure fell considerably ; the portal fall was very large, having fallen to the lower margin of the paper. Began infusion of nine hundred and fifty cubic centimetres of saline at 42° C. At this point respiration failed and artificial was supplied. Blood-pressure gradually increased during the entire time of infusion. Amplitude of heart-beats increased. The rise in the portal pressure, as shown by the water-manometer, was very marked. While changing the drums the heart's action became weak and the blood-pressure fell considerably. Infusion of one hundred and fifty cubic centimetres of saline produced a rise in all the manometers. Elevating the foot of the dog-board caused a considerable rise in all the pressures. Upon returning to its original position the central, cephalic, and peripheral pressures fell below their normal level—that is, the level just before the dog was elevated. The portal remained at a considerably higher level. During the time of raising the board, and following it, artificial respiration was discontinued ; voluntary respirations were sufficient. Intravenous injection of one-fiftieth of a grain of strychnine sulphate was followed at first by a slight rise in the central, then a slight fall in the central and cephalic, then a sharp rise in all the manometers. Respiration ceased at this juncture, and the frequency and amplitude of the heart-strokes increased. After forty seconds vigorous artificial respirations were made for twenty seconds, then discontinued for fifty seconds, during which time feeble, infrequent, and inefficient voluntary respiratory efforts were made. Artificial respiration was begun and continued throughout the experiment. Application of Bunsen's flame produced a sharp rise in central and cephalic, slight rise in peripheral, no appreciable effect in the portal. Bunsen's flame to the intestines was followed by a fall in the central, portal, and cephalic, and slight rise in peripheral. Intravenous injection of one-thirtieth of a grain of nitroglycerin was followed almost immediately by a slight rise in the central, then a sharp fall in the central, cephalic, and portal, but quite a marked rise in the peripheral. The frequency and amplitude of strokes were unchanged. Raising the head of the board was followed by an increase in all the areas. Artificial respiration was discontinued and the animal would not breathe. Heart-beat strong. Animal allowed to die of respiratory failure.

Post-Mortem.—The heart entirely empty in systole. The vena cava inferior was much engorged. The large radicals of the portal system quite full. The intestines, aside from those excluded in placing the portal canula, were not greatly congested. The stomach, duodenum, and descending colon rather pale. The veins of the extremities quite full. The liver large, hard, and congested, and on making an incision a large quantity of bluish watery blood escaped. The kidneys contained a comparatively small quantity of blood ; the spleen was very blue and moderately congested.

CXI

February 26, 1897.—Dog; weight, seventeen kilos. Central, cephalic, and portal tracings taken after the shock incident to the necessary operations. Administered saline solution quite hot (48° C.). Portal pressure was quite low at the time, but a prompt rise in all the pressures followed. Manipulation of the testicles produced a sharp fall in the central, peripheral, and cephalic pressures, also increased amplitude of the heart-stroke. The portal pressure alone rose. The respirations were very much increased in frequency. Forcible dilatation of the rectum was followed at first by a slight rise in central, cephalic, and peripheral, then a fall in same; the portal remained unchanged. Respirations increased in frequency and in depth. After these procedures the portal gradually fell to a low level. Saline produced a rise in all the pressures. Withdrawal of the intestines produced a marked rise in the portal pressure. There was no change in the other manometers.

CXII

February 27, 1897.—Dog; weight, fifteen kilos. While attaching all the canulæ the blood-pressure fell considerably and the cardiac action became weak. One hundred cubic centimetres of saline were introduced at the temperature of 43° C. into the external jugular vein. This was followed by a prompt rise in the central, cephalic, peripheral, and portal pressures. The amplitude of the heart-strokes increased and the respirations were deeper. Some time was spent in adjusting the apparatus; meanwhile the blood-pressure had fallen and the pulse-wave had become diminished. Three hundred and fifty cubic centimetres of saline at 42° C. caused a marked rise in central pressure and an increased amplitude in the strokes. A moderate increase in the other manometers was also noted. Artificial respirations were now discontinued, as feeble voluntary efforts were observed. Thirty seconds after the artificial respiration was stopped the central and the peripheral blood-pressures began to fall; the cephalic and portal were still rising. One minute after artificial respiration was discontinued the heart ceased beating. Artificial respiration was begun at once, but the heart did not again resume its action.

Post-Mortem.—The heart was empty in systole, excepting the right auricle, which contained a clot extending into the vena cava. Observation immediately after the thorax was opened showed that an occasional heart-beat still occurred. It was thought that a contractile movement could be seen, beginning in the walls of the superior vena cava and the vena azygos major, at a distance of one inch or one and a half inches from the mouths of these structures. All the great veins were much engorged with a dark watery blood, which clotted quickly when removed. The liver was extremely congested, large, and firm, quantities of blood issuing when incisions were made. The kidneys contained a comparatively small quantity of blood. The lungs were pale and almost bloodless.

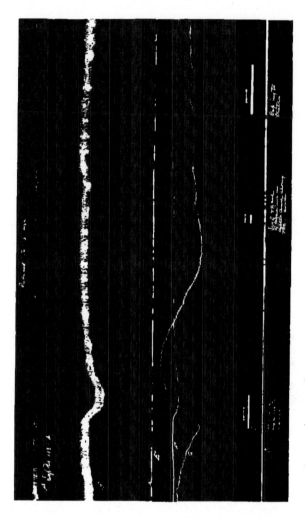

THE EFFECT OF INJURY TO TESTICLE AND ITS PREVENTION BY INJECTION OF COCAINE INTO THE SPERMATIC CORD.—A, respiration; B, central blood-pressure; E, venous peripheral in femoral vein; F, cephalic peripheral in carotid; G, portal splenic vein. Note effect of severe injury of testes, a sharp decline in the central and peripheral venous, and sharp rise in the portal, indicating a dilatation of the vessels of the splanchnic area. Injected four per cent. solution of cocaine into testes and along spermatic cord. Repeated more severely the first experiment, showing the value of cocaine in preventing shock on male genital organs, since even greater manipulation after cocaine injection produces no effect.

CXIII

February 28, 1897.—Dog; weight, fourteen kilos. Time of experiment, three hours and twenty-five minutes. The technique was carried out as in the preceding case; the portal canula trailed half an inch. Severe injury of the testicle produced a sharp fall in the central and cephalic, slight rise in the peripheral venous; the portal was falling at the time of the experiment. There was temporarily acceleration of the fall, followed by a rapid rise. Injected a four per cent. solution of cocaine into the other testicle and cord. No appreciable effect. Two minutes of continuous injury produced no change in the pressures, excepting the por-

INTRA-LARYNGEAL MANIPULATION; ELEVATION OF FOOT OF DOG-BOARD; REMOVAL OF INTESTINES.—A, respiration; B, central blood-pressure; C, signal; D, seconds; E, peripheral venous femoral; F, peripheral cephalic; G, portal. Note on the left the effect of intra-laryngeal manipulation, the marked rise in the portal pressure on elevating the foot of the dog-board. Note particularly on the extreme right the sharp ascent of the splanchnic manometer on exposing the intestine. The pressure sent the manometer up so rapidly and so far as to cause a marked overflow, thereby preventing the complete tracing of the actual height. This tends to show that the splanchnic vessels dilate on exposing the intestine, and so produce a fall in blood-pressure.

tal, which showed a slight gradual rise. After one and a half minutes, forcible dilatation of the rectum was followed by a fall in the central and cephalic, a slight rise in the portal, and slight irregularity in respiration. After this respiration gradually failed, becoming more infrequent, and displaying a peculiar jerky inspiratory and expiratory character. During this period of gradually failing respiration the central pressure rose slowly, the vasomotor curves became more prominent. During the same period the portal pressure suffered a very great fall. Artificial respiration at this point produced a slight fall in the central cephalic, a corresponding rise in the peripheral venous ; portal effect not registered. Injection of fifty cubic centimetres of saline produced sharp rise in the central and portal, a slight rise in the cephalic and peripheral. Voluntary respirations, which had

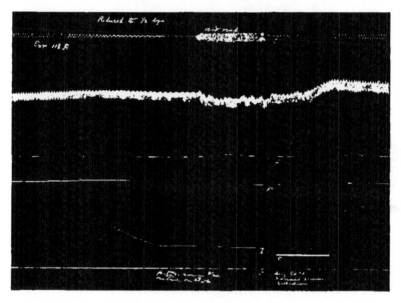

SHOWING EFFECT OF ARTIFICIAL RESPIRATION AND OF INFUSION OF SALINE SOLUTION.—*A*, respiration; *B*, central blood-pressure; *C*, signal; *D*, seconds; *E*, peripheral venous femoral; *F*, cephalic; *G*, portal. Note gradual failure of respiration up to the point at which artificial respiration was supplied. There was also beginning failure of the circulation, as evidenced by longer vasomotor curves. The infusion of fifty cubic centimetres of saline caused upward tendency of all pressures, especially the central and the portal. Artificial respiration was followed by gradual recovery of the respiratory rhythm.

been resumed, were much improved. Bunsen's flame applied to the posterior and anterior extremities produced a marked rise in pressure in all the areas. Portal pressure followed central closely. Bunsen's flame to the intestines produced a gradual fall in the central, cephalic, and portal. The peripheral venous showed at first a gradual rise, then a gradual decline. Bunsen's flame over region of heart produced a gradual rise in the central, portal, cephalic, and peripheral ; respirations a little more frequent and deeper. Severe intra-laryngeal irritation was followed by an immediate and almost complete cessation of respiration, and very soon

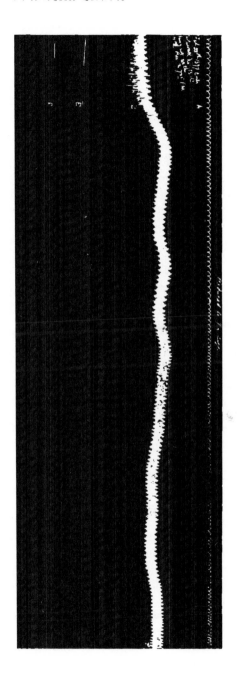

the character of cardiac action changed. Long, sweeping, slow beats appeared. Central and portal fell sharply; cephalic, a moderate fall; peripheral venous, very slight. Upon cessation of stimulus, respirations, though a little less frequent, rapidly regained normal action. Central and portal pressures rose rapidly; frequency of pulse increased. Elevating foot of dog-board was followed by considerable rise in the portal, slight rise in the central. Incision of abdomen was followed by fall of the central. Withdrawal of the intestines was followed by great rise in the portal, very little effect on the central; a slight fall. After intestines were returned the portal continued at a high level for some time. Tracing *F* shows the effect of twelve hundred cubic centimetres of saline, but pressure was low, pulse weak, respiration shallow, all of which were improved. Animal was killed by forcing air into his jugular vein. Heart's action became slightly irregular, then suddenly stopped.

Post-Mortem.—Left chambers empty and right chambers of heart contained blood and air. A small clot in the right auricle. The large venous trunks were considerably engorged, and the vena cava inferior contained a quantity of air. Liver much engorged and contained a large quantity of bluish, watery blood. Kidneys contained comparatively less blood than the liver, but seemed to contain more than those of other dogs under similar conditions. The loops of intestines from which the circulation had been cut off seemed quite blue and were much engorged. Other portions presented nearly normal post-mortem appearances.

CXIV

March 1, 1897.—Dog; weight, twenty-five kilos. Time of experiment, three hours and fifteen minutes. Forcible dilatation of the rectum; rise in central and peripheral; portal, cephalic, and venæ cavæ unchanged. The vena cava inferior was connected with a water manometer by passing the slender canula through the portal vein into the vena cava. The canula was inserted from six to eight inches. Incision of skin over testicle produced no effect. Severe injury of testicle produced a slight fall in central pressure; no change elsewhere. Incision of other side of testicle, no effect. Severe injury to left, slight fall, and then, towards termination of the stimulus, a rather sharp rise in central and peripheral. Respirations less deep and full than during the period of stimulation. Tracing *B* was begun just after respirations failed. No experiment was made during the first part of *B*. One hundred and twenty-five cubic centimetres of normal saline injected; increase in central. After about one hundred cubic centimetres of saline, there were several long, sweeping heart-strokes, which were repeated after an interval of five or six seconds. After a few such strokes the heart's action became regular. The blood-pressure remained at the level to which it had been raised. Portal, peripheral, and cephalic were increased, but were irregular, in harmony with the central irregularity. The beginning of Tracing C shows natural respiration, regular, but rather infrequent and spasmodic in character. All the pressures pursued a uniform course. A small quantity of saline produced an upward tendency in all. By mistake a small quantity of saturated solution of sulphate of magnesia entered the portal circulation; sharp rise in central, followed by a marked and rapid fall, then sudden stopping of the heart and respiration at the same instant. Vigorous artificial respirations were of no avail. Exact amount of solution could not be determined, but it certainly was very small.

Post-Mortem.—Heart empty in systole; small clot in right auricle; venous trunks not so much engorged as in previous cases, in which death was from shock alone. The intestinal loops appeared paler than normal Liver not so much engorged.

SHOWING TOXIC EFFECT OF IN-TRAVENOUS INJECTION OF MAGNE-SIUM SULPHATE.—*A*, respiration; *B*, blood-pressure; *E*, peripheral venous; *G*, portal. Magnesium sulphate, saturated solution, accidentally admitted into portal vein. Note sudden death.

CXV

March 3, 1897.—Bitch, old; weight, nine kilos. Time of experiment, one hour and twenty minutes. *First.* Sharp blow upon epigastrium with a small hammer; very little effect. Slight fall in the central. *Second.* Severe dilatation of the rectum; slight temporary increase in the peripheral, venous, and the portal pressures; others unchanged. *Third.* Severe dilatation of vagina was followed by similar phenomena. *Fourth.* Injection of fifty cubic centimetres of saline, temperature of 42° C.; increase in central, portal, and peripheral, and a decided improvement in respiration. One hundred cubic centimetres of saline additional; further increase in pressure and improvement in respiration. Shot the animal with a .32-caliber Smith & Wesson revolver. Respiration ceased at once; heart-strokes began to diminish in amplitude; blood-pressure rose for a few seconds, then declined. Heart continued beating for nearly two minutes after respiration had ceased.

Post-Mortem.—Heart in diastole. Clots in the right and in the left auricle and in the right ventricle, the left ventricle nearly empty, lungs blue and considerably engorged, large veins not very full and vessel of the splanchnic normal. The course of the ball was as follows: Entering the anterior aspect of the neck, passing the thyro-hyoid space just to the

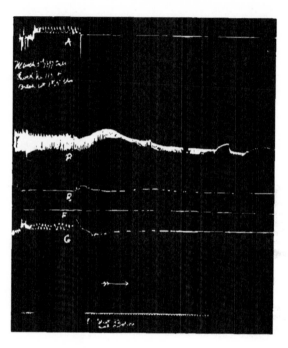

GUNSHOT WOUND.—A, respiration; B, central blood-pressure; E, peripheral venous in femoral; F, peripheral cephalic; G, portal. Shot passed through neck, cutting a small notch in the medulla. Note sudden cessation of respiration and continuation of the heart-beat; also note the rise in blood-pressure, probably in consequence of arrested respiration.

right side of the median line, thence through the epiglottis, the posterior wall of the pharynx, passed backward between the atlas and the occiput, cutting a notch in the condyloid process of the latter, and cutting through and destroying a small portion of the right side of the medulla oblongata, the ball passing on backward and emerging just to the right side of the median line.

CXVI

March 3, 1897.—Small brown mongrel pup; weight, five kilos. The animal took the anaesthetic badly, and the respiration and the heart failed soon after beginning the experiment.

CXVII

March 3, 1897.—Puppy about five months old; weight, six kilos. A poor subject. Respiratory failure occurred early in the experiment. Artificial respiration. Heart-action was weak and irregular from the beginning. Injection of saline solution produced some increase in blood-pressure and a little improvement in the character of the heart's action. The heart soon became more irregular, then failed.

CXVIII

March 4, 1897.—Dog; weight, twenty-three kilos. Time of experiment, two hours and fifteen minutes. After the canulæ were introduced the animal was in good condition. Severe injury to, and destruction of, the structures of the left tympanum; fall in central, portal, and cephalic pressure, and an increase in peripheral venous. A similar injury upon the opposite side; result similar to the above, excepting a slight fall in the peripheral venous. Following the fall a gradual rise occurred. Severe injury to nose, breaking the turbinated bodies and injuring the mucous membrane; diminished frequency of respiration, and immediately after the cessation a sharp, slight fall in central and cephalic pressures, portal showed a rise, peripheral venous unchanged in a slight gradual decline. For some time after this the central, portal, and cephalic showed large, wave-like undulations; the testicle was incised and severe injury inflicted. The undulations above referred to obscured the result. The portal showed a marked rise, following a moderately rapid decline; other areas not appreciably affected. After some moments severe injury of the opposite side was done; the central first showed a slight fall, then a pronounced rise, followed by a gradual fall. Peripheral venous and cephalic rose towards the termination of the stimulus. Upon the portal the tendency was to check a gradual fall which was taking place at the time. Respirations now rapidly failed just after the femoral canula was inserted into the vena cava; blood-pressure rapidly fell; the heart-action slow and irregular. Artificial respirations were of no avail.

Post-Mortem.—A clot of considerable size in right auricle; other chambers of the heart were practically empty. The large venous trunks moderately engorged and imparted a sense of considerable tension, especially in the inferior vena cava. Kidneys normal; liver somewhat congested; spleen congested; lungs slightly paler than normal, possibly due to the persistent artificial respiration after death had occurred.

CXIX

March 5, 1897.—Pug-dog; weight, twenty-nine kilos. Time of experiment, two hours and five minutes. Central in carotid, cephalic in peripheral carotid, peripheral venous in femoral vein, portal and inferior vena cava; canulæ introduced. Portal and inferior vena cava connected with water-manometers as usual. Respiration ceased, artificial supplied. Injected four hundred and fifty cubic centimetres saline; temperature 42° C. A general upward tendency, the vena cava more prompt than

EFFECT OF INTRAVENOUS SALINE SOLUTION.—A, respiration (artificial); B, central blood-pressure; C, signal; D, seconds; E, peripheral venous pressure; F, peripheral cephalic arterial pressure; G, portal canulæ pointing towards heart; H, vena cava inferior, canulæ pointing towards heart. Observe effect of the infusion of four hundred and fifty cubic centimetres of saline. Artificial respiration throughout.

the other pressures; likewise its fall was more rapid on discontinuing the saline. The pressure in the other areas was well sustained at the level to which the saline had raised them. Vasomotor curves quite pronounced after four hundred cubic centimetres of saline had been given. Injection of fifty cubic centimetres produced similar results, but in a less degree. While the drums were being changed there was considerable fall in the general pressure, but the fall in the vena cava pressure was most marked. Injection of two hundred cubic centimetres, temperature of 42° C.; rise in all the areas, that of the vena cava very rapid. While removing a clot from the cannula artificial respiration was not well done, and blood-pressure rose six millimetres. Upon resuming competent artificial respiration blood-pressure fell sharply fifteen millimetres. Injury to structure of

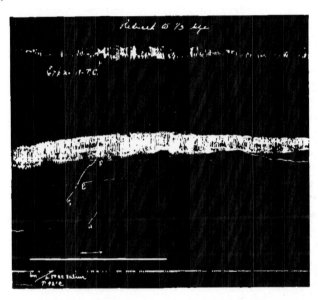

EFFECT OF INTRAVENOUS SALINE INFUSION.—*A*, respiration (artificial); *B*, central blood-pressure; *C*, peripheral venous in femoral; *F*, cephalic peripheral, carotid artery; *G*, splanchnic in splenic vein, cannula pointing towards the liver,—the remainder of the splanchnic area intact; *H*, vena cava inferior, long cannula inserted into femoral, passed up towards the heart, and taking the pressure in trunk of the vessel; cannula so small as to not materially interfere with the venous flow. White line at bottom of cut indicates the period during which two hundred and fifty cubic centimetres of saline solution were injected. Respiration failed early in the experiment, artificial supplied. Four hundred and seventy cubic centimetres had been previously given. The peripheral venous paralleled the central. The vena cava rose promptly and likewise fell. The portal rose later, and was partially sustained. The evenness of the curve at a comparatively low level is indicative of greatly impaired vasomotor tone of the vessels. While there is a general upward tendency of the central pressure after the beginning of the flow, this tendency is much less than a like amount would produce in an animal of equal size whose vasomotor centre had not been so much impaired. The respirations are artificial. The animal died without re-establishing natural breathing.

right tympanic cavity; fall in central and peripheral pressures, and a slight rise in the portal. Vena cava doubtful, owing to unfavorable conditions in its registering. Injury to structure of left tympanum; rise in central, slight fall in peripheral; portal clotted. Inferior vena cava was falling rapidly at the time. Fifty cubic centimetres additional; increase in blood-pressure and in amplitude of the stroke. Injection of two hundred and fifty cubic centimetres; considerable rise in all the manometers, and well sustained in central, but the pressure began to decline in other regions immediately upon cessation of the flow of the saline. Opening the abdomen caused a slight fall in the central, peripheral, and portal. Pulse-wave was diminished. On withdrawal of the intestines there was a fall in the blood-pressure in the central, and relatively greater in the portal than in any other manometer. After some moments two hundred cubic centimetres of saline introduced; slight rise in central, increase in other regions more tardy. Portal and inferior vena cava did not indicate a change until fully one hundred cubic centimetres had been introduced. The character of the pulse was not improved, but continued rather to deteriorate. The mechanism was stopped to effect some adjustments, and in this time cardiac action ceased. Artificial respiration was maintained throughout the greater part of the experiment.

Post-Mortem.—The left auricle and both ventricles were comparatively empty. Small clot and a small quantity of blood in the right ventricle; a clot of greater size in the right auricle; the intestine somewhat paler than usual. It was thought the mesenteric arteries were dilated. The liver was moderately congested. Watery blood flowed freely upon incision. The large venous trunks were engorged and tense; the kidneys very slightly, if at all, congested; the lungs apparently normal.

CXX

March 6, 1897.—Bitch; weight, twenty kilos. Death occurred just after portal and inferior vena cava tubes were released—that is, when the pinch-cocks were opened. The animal had been doing well up to this point, and it was at first thought the anæsthesia was incomplete. A peculiar straightening and convulsive movement of the limbs was noted. Immediately thereafter respiration ceased, and subsequently the heart rapidly failed. The individual heart-strokes were strong and full, but became more and more infrequent and irregular, and soon ceased.

Post-Mortem.—Heart in systole; blood and clots in both right chambers and in the left auricle; left ventricle empty; splanchnic veins very full; splanchnic arteries empty; liver intensely congested; large venous trunks greatly engorged.

Note.—The portal and the inferior vena cava canulæ were directed towards the heart; the fluid in the pressure-bottle was a saturated solution of sulphate of magnesia. If the pressure of the solution filling the rubber tubing were greater than that in the blood-vessels, the solution would flow directly into the circulation; if it were less than the blood-pressure, blood would flow into the tubing to equalize the pressures.

CXXI

March 6, 1897.—Dog; weight, fourteen kilos. In precisely the same manner, and at the same point in the experiment and under similar circumstances, this animal died. The post-mortem appearances were identical. NOTE.—The solution obviously flowed into the vessels.

CXXII

March 7, 1897.—Large, strong dog, had been safely anæsthetized. The experiment was almost an exact duplicate of the preceding. The post-mortem appearances were practically the same.

CXXIII

March 6, 1897.—Young spaniel; weight, ten kilos. Respirations failed early in the experiment; artificial were supplied. *First.* Injection of one-fifteenth of a grain of strychnine into the jugular vein was followed after fifteen seconds by a very marked and rapid rise in blood-pressure in all the manometers, but greatest in the central. The pressures were well sustained and continued high for some time. *Second.* Skin and muscles of abdomen incised; fall in all the pressures. *Third.* Peritoneum incised—central pressure was rising at the time; no immediate effects, but soon after cephalic followed the central. Very little change in the peripheral venous. Portal showed at first a fall and then a subsequent sharp rise, not noted in the others. *Fourth.* Manipulation of the intestines; gradual fall in the central and cephalic; no change in the peripheral venous, slow rise in the portal. *Fifth.* A subsequent manipulation; slight gradual fall in all pressures, with slight general increase upon cessation of the stimulus. *Sixth.* The procedure repeated; general fall. On returning the intestines, there was gradual increase in the portal, and slight gradual decline in the other pressures. *Seventh.* Severe injury causing partial destruction of the right tympanum; increase in central and peripheral venous, decline in cephalic; no change in portal; pulse small. Three hundred and fifty cubic centimetres of saline injected; general rise. Artificial respiration had been continued up to this point, but it was now discontinued. There was an immediate but slight rise in blood-pressure, followed by a slight but sharp fall; then some rise for a period of one minute, after which there were variations of a similar kind. After a number of variations the heart stopped suddenly. The portal curve followed the central closely; cephalic did also in the main. The peripheral venous rose slightly upon cessation of artificial respiration, and then began to decline much earlier than did the other pressures. Vigorous artificial respiration, following the failure of the heart, was attended by a single beat, which was not repeated.

Post-Mortem.—Left ventricle empty ; some blood in the other chambers. Liver hard and tense, much engorged. All the large venous trunks engorged ; lungs, normal ; kidneys, normal.

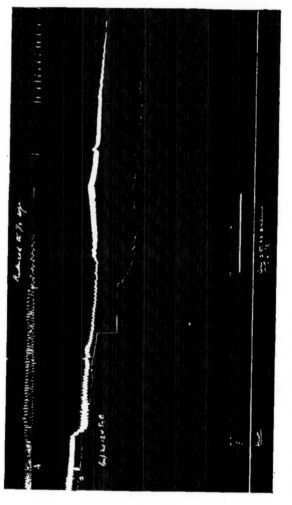

GUNSHOT TRACEY.—*A*, respiration ; *B*, blood-pressure ; *C*, shot fired ; *G*, portal pressure. In the central blood-pressure tracing the individual manometer excursions blend on account of the rapidity.

CXXIV

March 9, 1897.—Dog ; weight, nineteen kilos. Time of experiment, two hours and eighteen minutes. Vena cava manometer did not correctly register in the first tracing. Injection of two hundred cubic centimetres of saline at 41° C. was attended by a rise in the central and the peripheral ; the portal and the vena cava were not correctly registering. Respi-

rations much improved. Withdrew omentum; no effect. Manipulation
of the mesentery and withdrawing intestines; slight fall in central and a
very sudden fall in portal, the latter partly due, it is believed, to mechanical causes. Respirations failed; rise and then fall in blood-pressure.
Artificial respiration begun and for a time of no avail, but at length the
heart began beating again. Blood-pressure became quite high and the
pulse much improved. On resuming voluntary respiration the blood-
pressure declined. The intestines were returned; slight fall, then a higher
rise in the portal. A shot was fired into the animal's head from a 32-
caliber revolver. The ball entered the anterior surface of the neck, passing through the thyro-hyoidean space one and a half inches to the left of
the median line, passed through the base of the epiglottis, backward
through the post-pharyngeal wall, through the muscular masses between
the occiput and the left transverse process of the atlas, deflected slightly
outward, and escaped through the muscular and the tegumentary tissues
of the neck; the cord was not lacerated. The immediate effect was a
marked slowing of respiration, a sharp but comparatively small fall in
blood-pressure; at first increase, then decrease in the amplitude of the
pulse-waves; respirations gradually slowed, became irregular, then failed.
During this time blood-pressure fell slowly after respirations had well-
nigh ceased; the heart-strokes gradually became greater and the central
and portal lines began a gradual rise, which continued for about one
minute and then commenced to fall slowly. This rise seemed to be due
to the introduction of two hundred and fifty cubic centimetres of saline;
very soon after this the respiratory effort was noted. After about forty
seconds other respiratory efforts occurred at intervals of from eight to ten
seconds. This continued for a period of about one and a half minutes,
during which time cardiac action gradually became weaker and failed
soon after entire cessation of respiration.

Post-Mortem.—Heart in diastole. Blood in all the chambers, but
little in the left, however. Great veins moderately full; liver much
engorged, hard, and cyanotic. A large quantity of blood flowed on
making incision. The kidneys were congested more than in most of the
preceding experiments. Moderate distention of the veins of the mesentery; mesentery arteries not dilated.

CXXV

March 10, 1897.—Dog; weight, seventeen kilos. Rather young; long,
woolly hair. Time of experiment, two hours and fifteen minutes. Experiments as before. *First.* Severe injury to, and partial destruction of,
the structures of the left middle and internal ear; sharp rise in central,
then one long, sweeping down-stroke, then another slight, sharp rise,

then a sharp decline, and, subsequent to the period of stimulation, a
gradual increase in blood-pressure. Peripheral venous showed a sharp
rise and quick return to former level; was due to a convulsive, light
movement of the right hind limb, which was connected to the canula.
The portal showed a sharp rise, followed by a short period of more
gradual ascent, and then by gradual fall. The portal followed the rise
in central, subsequent to the period of stimulation, more closely than did
the peripheral. Respirations were hurried and irregular during the stimulation, becoming slower and more shallow for a short time afterwards,

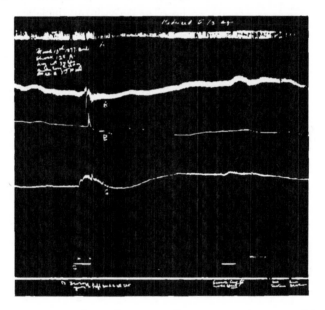

INJURIES INVOLVING THE MIDDLE EAR AND THE NARES.—A, respiration; B, central blood-pressure; E, peripheral venous in femoral; G, portal. Note on the left the effect of injury of the
structures of the middle ear. Note the effect of injury of the nose: nares contused, turbinated,
crushed.

then they returned to their normal character. *Second.* Severe injury to
the structures of the nares; breaking up the turbinated bones, etc.; was
followed by a slight, gradual rise and then corresponding fall in all the
pressures. *Third.* Injury to the testicle by repeated sharp, light blows;
was followed by a very abrupt fall in central and by a moderate rise in
the peripheral and portal. Respiration became quite shallow. On cessation the tracing became quite normal. A repetition of this experiment
brought out like phenomena. *Fourth.* Injections of one hundred cubic
centimetres saline solution at 41° C. Considerable general rise in blood-
pressure and improvement in respiration. *Fifth.* No. 3 repeated with

like results. Infusion of one hundred cubic centimetres of saline as before. *Sixth*. Shot from a .32-caliber revolver was fired into the animal's head; the ball entered just internal, and slightly anterior to the angle of the inferior maxilla, ranged slightly forward and inward, traversed the muscular structures of the floor of the mouth, the base of the tongue, the oral vault, entering the left posterior nares, puncturing the palate bone, crushing through the body of the sphenoid, shattering this bone and the ethmoid, causing extensive laceration of the brain-tissues, meninges, and vessels of this region; thence the ball passed

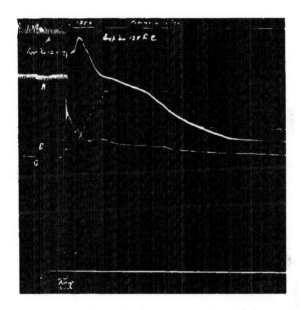

GUNSHOT INJURY.—*A*, respiration; *B*, blood-pressure, central; *D*, signal; *E*, peripheral venous, femoral; *G*, portal. For the course of the ball see protocol. Note respiratory arrest. The central blood-pressure shows merely a white line; the strokes were so rapid as to make their reproduction in the cut impossible, indicating a loss of the inhibitory influence, with probable stimulation of the accelerator.

through the left cerebral mass and lodged just beneath the vault of the cranium. This portion of the skull was extensively fractured. The bullet was badly scarred, and the bits of bone which it detached during its passage through the sphenoid caused very extensive lacerations of the brain-tissues. The immediate result was cessation of respiration, a marked increase in the central and the portal pressures, the latter likely due to mechanical causes—viz., the convulsive fixation of the muscles of the abdominal wall. Peripheral venous rose sharply also, likely due to the convulsive contraction, and then fell rapidly to its former level. The sharp rise in central pressure was soon followed by a marked fall, and then by a period of moderately rapid decline until the abscissa was reached. The pulsations were infrequent at the end. Not a single respiratory movement was noted after the shot.

Post-Mortem.—Heart in diastole; blood in all the chambers; least in the left ventricle; lungs normal; large veins moderately filled; liver engorged; kidneys moderately so; veins of mesentery fairly full.

CXXVI

March 10, 1897.—Dog. Time of experiment, two hours and fifteen minutes. Central, peripheral, portal, and inferior vena cava pressures taken. *First.* Incision of the skin of the abdominal wall; fall in central and inferior vena cava and a rise in the portal. *Second.* Withdrawal of the intestines was followed by a slight rise, followed by a fall in all the

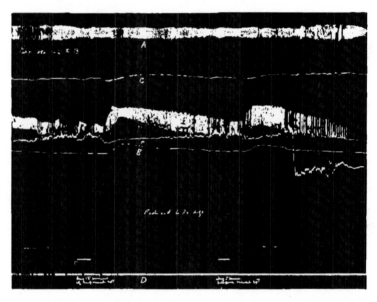

INTRAVENOUS INJECTION OF SUPRA-RENAL EXTRACT.—*A*, respiration; *B*, central blood-pressure; *C*, signal; *D*, seconds; *E*, peripheral venous; *H*, vena cava inferior. Canula towards heart. Note the effect of intravenous injection of supra-renal extract.

pressures. *Third.* Manipulation of the intestines; fall in the central and in the peripheral venous; inferior vena cava declined during the period of actual manipulation and rose slightly at its close, and after the manipulation it was at about the level it occupied before. Portal remained practically unchanged. *Fourth.* Whipping the intestines; slight fall, then subsequent rise in central. Pressure in the portal and the inferior vena cava decreased during the whipping. Following this there was a rise in the central. *Fifth.* Manipulation of the parietal peritoneum ; fall in the central and the portal ; no appreciable change elsewhere. After this experiment respirations failed and central blood-pressure fell. Artificial respirations were attended by a further fall in the central pressure and a slight rise in the portal and inferior vena cava. After some seconds the heart's action became irregular. *Sixth.* Injection of fifteen minims of supra-renal extract was followed by a slight increase in central and a slight fall in the inferior vena cava. Almost immediately after administration, after a few seconds of slow and irregular heart-action, during which the central pressure fell slightly and the vena cava and portal rose, there occurred a sharp increase in the central pressure ; at first a fall in the portal and inferior vena cava, subsequently a rise in the portal, vena cava, and peripheral venous. In these pressures there was a continued rise after the central had begun to fall. After two minutes' delay five minims were injected, with results similar to the above. The action of the heart soon became irregular and infrequent and ceased.

Post-Mortem.—Heart in systole ; clot and small quantity of blood in the right auricle ; other chambers of the heart empty. Great vessels moderately engorged ; the liver likewise ; kidneys and lungs practically normal.

CXXVII

March 11, 1897.—Bitch ; weight, twenty-two kilos. Time of experiment, two hours and five minutes. Central, peripheral, vena cava, and portal pressures recorded. Portal imperfect in first tracing. *First.* Dilatation of the anus and the rectum ; rise in pressure. *Second.* Effect of abdominal section obscured. *Third.* Four hundred cubic centimetres of saline at 40° C. ; usual results. *Fourth.* Manipulation of uterus ; rise in central ; no effect in the other pressures. *Fifth.* Manipulation of the ovaries and the tubes during the progress of the general decline of blood-pressure, mentioned above, arrested the fall and caused slight rise. A slight increase in the other pressures was also noticed. On cessation of the stimulus the pressures fell. Respiration was not appreciably affected. It became shallow and slightly irregular at the beginning of the manipulation. *Sixth.* Withdrawal of the intestines was followed by a gradual decrease of blood-pressure. *Seventh.* Return of intestines to the cavity ; slight rise in central and peripheral ; a more marked rise in the inferior vena cava. The portal did not register any effects. *Eighth.* Injection of five minims of supra-renal extract ; a sharp rise in central and peripheral was noted. Inferior vena cava at first fell slightly, and then followed the general rise. After the sharp rise in the central there was an immediate and abrupt

fall to a point about twelve millimetres above the level occupied before
the injection.   Following the injection there was a slight and well-sus-
tained acceleration of the heart's beat; rhythm and force practically un-

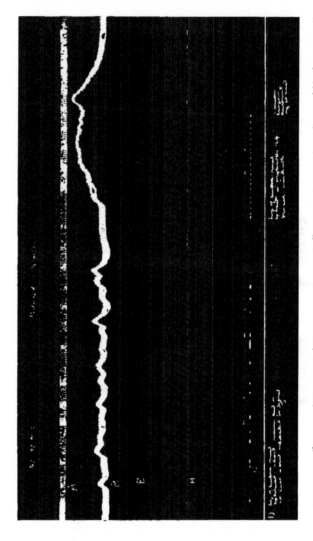

SHOWING THE EFFECT OF INTRAVENOUS INJECTION OF SUPRA-RENAL EXTRACT.—*A*, respirations; *B*, central blood-pressure; *E*, peripheral venous; *H*, vena cava inferior.   Dots indicate intravenous injections of the suprarenal extract.   A larger amount was injected at last.

changed.   *Ninth.* Injection of a small amount of the supra-renal extract
at comparatively short intervals, varying from two to ten seconds, was
followed in each instance by a slight, but prompt, rise in the pressure.

About one minim of the extract was injected at each period marked. On the whole, after the injection had been given, there was a slight gain in the general pressure. Ninety minims more were injected at short intervals, one minim at each injection; each small dose showed its distinct, individual effect. In this way, before the pressure began to fall, a new increment was added, and consequently the pressure was forced up to a considerable height. However, a gradual decline followed. There was but little permanent gain. No effect upon respiration was observed, and the heart's action continued strong and full afterwards. *Tenth.* Shot from a .32-caliber revolver was fired through the animal's larynx and the soft tissues of the neck; there was quite a sharp hemorrhage, and the blood-pressure fell rapidly. Directly afterwards it became steady, respirations were arrested for an instant and then resumed with diminished frequency. A second shot was now fired, which entered below the inferior maxilla, a short distance internal and anterior to the angle of the jaw. It passed upward, slightly backward and towards the median line, passing through the left cerebral mass, lacerating the basal ganglia. The body of the sphenoid bone was extensively fractured, opening the lateral sinus and the left lateral ventricle. The left side of the cranium was also extensively fractured. This shot was followed by immediate cessation of respiration and a slight fall in blood-pressure, the character of the heart's action remaining unchanged. The blood-pressure continued for some time fairly constant. Just before the heart failed one hundred and fifty cubic centimetres of saline were injected and artificial respirations were supplied, but it was of no avail.

CXXVIII

March 12, 1897.—Dog; weight, eighteen kilos. The animal did not take the anæsthetic well, and a part of the experiment was made under incomplete anæsthesia. *First.* Dilatation of the anus and the rectum produced a sharp fall in the central, rise in the portal and inferior vena cava, due probably to the contraction of the abdominal muscles on account of incomplete anæsthesia. The pressure was so irregular at this time that the effect was much obscured. (See Tracing *A.*) *Second.* Under complete anæsthesia, Bunsen's flame applied to the skin caused a sharp, considerable rise in central and portal and a moderate decrease in the peripheral. The depth of respirations was increased. *Third.* Injury to the middle ear was followed by a gradual fall in the central and a rise in the portal, also a slight increase in the inferior vena cava. *Fourth.* Intra-laryngeal manipulation produced absolute cessation of cardiac and respiratory action; a sharp fall in central and in portal, the inferior vena cava considerable, peripheral venous but little changed, a slight fall. On cessation of the manipulation the central pressure returned to its former level. Then a series of deep, irregular beats occurred, during which there was a fall. Then the heart-action resumed its former character, and blood-pressure rose to former level. The respiration, on cessation of the stimulus, began after an interval, was infrequent and irregular, but soon resumed its nor-

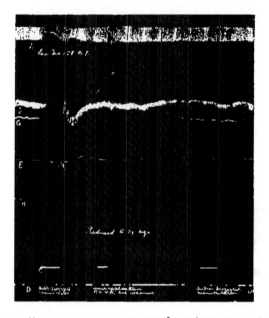

INTRA-LARYNGEAL MANIPULATION BEFORE AND AFTER LOCAL APPLICATION OF COCAINE.—A, respiration; B, central blood-pressure; G, portal pressure. Showing effect of intra-laryngeal manipulation on blood-pressure and respiration. The blank space in the blood-pressure tracing shows in original very striking fall in the pressure parallel to the drop in the portal. In the second experiment cocaine had been applied. Note complete arrest of respiration during intra-laryngeal manipulation, as shown above the first signal on the left. The middle signal designates the local application of four per cent. solution of muriate of cocaine upon the laryngeal mucosa. Note slight inhibition of heart and temporary arrest of respiration during its application. The signal on the right marks the application of severe intra-laryngeal manipulation. Note the total absence of changes in the tracing.

mal rhythm, frequency, and depth. Cocaine was now applied locally to the laryngeal mucosa. *Fifth.* After about one hundred seconds similar manipulation was made, with no other effect than a slight fall in the central pressure. *Sixth.* After some moments, administered a small quantity of supra-renal extract by intravenous injection; was followed by the same phenomena noted after a similar procedure in CXXVII—sharp rise, followed by fall.

CXXIX

March 13, 1897.—Brown-haired dog; weight, nine kilos. Time of experiment, one hour and forty-five minutes. Central, peripheral, and portal pressures taken. While connecting the portal manometer a small quantity of fluid from the pressure-bottle was accidentally admitted into the splenic vein. The animal directly straightened out and exhibited convulsive movements. Respirations failed as the tambour was being attached. The heart soon failed and the blood-pressure fell. The portal blood-pressure gradually increased from the time the convulsive-like movements were observed. Artificial respirations were begun and maintained for about thirty seconds before any heart-action was noted. The markings on the tracings were due to mechanical causes incident to artificial respiration. After the above-mentioned period the heart began to beat, at first irregularly and infrequently, then the amplitude of the strokes increased, the blood-pressure rose. The heart's action became no more frequent and the amplitude of the strokes increased. The blood-pressure continued to rise until it had passed its former level. At this point artificial respiration was discontinued, as the animal showed beginning natural efforts. Then the following experiments were performed : *First.* Insertion of the glass canula into the urethra ; produced no effect except to slightly increase the respiratory movements. Blood-pressure was increased slightly at the time. Dilatation of the bladder by forcing fluid through the urethra produced a slight fall in the central pressure and rise in the portal and the peripheral venous. The respirations were slightly increased in depth, but the rhythm remained unchanged. The above procedure seemed to have the effect of abolishing or greatly disturbing the vasomotor curves, which appeared quite distinct both before and after this procedure. *Second.* The bladder having been dilated by forcing water into the urethra, the penis was clamped so as to occlude the urethra. Sufficient compression was now applied to cause rupture of the bladder. The line of rupture was parallel with the long axis of the organ, and upon the right side. The compression was followed by a rise in the central pressure and also in the portal. Blood-pressure declined after rupture, and the vasomotor curves became more marked. The respirations were more shallow during the period of compression. *Third.* Compressing testes, without injuring scrotum, produced no appreciable change. Respirations soon began to fail rapidly ; blood-pressure at first declined, then rose, and amplitude of the heart-strokes increased, after which the pressure began to fall ; the heart-strokes increased in length and became more irregular. Soon the heart's action became less frequent and the strokes diminished in length. Artificial respiration was begun ; slow heart-action and low blood-pressure continued for about twenty-five seconds, then the heart-action became more rapid and the strokes increased in length. Following this, the heart action became still more rapid, and the strokes began to diminish in length and the blood-pressure to rise. The cardiac action and the blood-pressure soon returned to the condition noted before the respiratory failure. After a few more moments voluntary respirations were resumed. A shot was fired from a .32-caliber revolver into the animal's head. The ball entered the floor of the mouth nearly at a point where a line drawn from the angles of the inferior maxillae would cross the median line. The ball passed through the tissues of the floor of the mouth, through the base of the tongue, through the base of the soft palate, the body of the sphenoid bone, and through the left cerebral mass near the median line. The ball lay loosely in the brain-tissue, just beneath a large button of bone, oval, seven-eighths of an inch by one and a quarter inches, which had been broken out of the cranial wall and slightly elevated. The lateral sinus had been opened by the injury, and a large hæmatoma was beneath the scalp. Respirations ceased immediately after the shot was fired. All the pressures rose sharply, then declined rapidly for a short time, and then pursued a fairly level and uniform course until the heart ceased to beat.

Post-Mortem.—Heart in diastole; blood in all the chambers; great veins but slightly engorged; intestines and stomach pale; liver contained less blood than normal; kidneys considerably engorged; lungs normal.

CXXX

March 13, 1897.—Dog; weight, fourteen kilos. Time of experiment, two hours and twenty minutes. Intra-laryngeal irritation; usual effect. This was made for a control. Injection of four per cent. solution of cocaine into the sheaths of the vagi. Intra-laryngeal manipulation; cessation of respiration during time of manipulation; no effect on blood-pressure or heart's action. The blood-pressure gradually increased after the cocaine was administered. The respiratory curves were very marked. Electrical stimulation. Used the faradic current, one Daniel cell, Du Bois-Reymond apparatus, secondary coil lapping the primary half an inch. Stimulation was applied just above the cocaine block, and produced a very great and very rapid rise of the blood-pressure. Remained at the elevated level for a few seconds, and then fell, on closing the key, as rapidly as it rose. The phenomenon seemed to be due to a powerful augmentative effect. The force of the heart's action was much increased; the rate was unchanged in part of the beats, and somewhat slowed in other parts. Repeated the stimuli four times, with like results. During the ascent in the second stimulation of the nerve there appeared a series of long, sweeping strokes. Electrical stimulation after a lapse of ten minutes produced practically no effect. Stimulation below the cocainized part stopped the heart instantly. Respiration continued; the heart did not recover.

Post-Mortem.—Heart in diastole; blood in the cavities; great veins and the liver considerably engorged; lungs somewhat congested; arteries of the splanchnic area empty; the veins moderately full.

CXXXI

March 15, 1897.—Dog; weight, nine kilos. Time of experiment, one hour. Bunsen's flame to the right hind foot was followed by a rather marked rise in central blood-pressure, by a slight rise in the portal, peripheral venous, and peripheral arterial. This was secured as a control. Injection of four per cent. solution of cocaine into the sheath of the sciatic and the anterior crural nerves. Bunsen's flame to the foot after the injection of the cocaine was followed by a slight and more gradual rise in the blood-pressure; the rise was delayed. Injection of one-fifteenth of a grain of strychnine to sustain the failing animal. There was an immediate fall in the central and a rise in the peripheral blood-pressures. Injection of saline solution into the nerve-sheath of the other limb, then Bunsen's flame applied, was followed by a slight fall in blood-pressure; however, the effect observed was not pronounced. Subsequent application of Bunsen's flame produced slight and rather contradictory effects. Hammer blows over parietal region of skull produced a depressed fracture, completely stopping respiration. At first the blood-pressure fell sharply, then continued to fall gradually and steadily. After about two minutes, and just as the heart was about to fail, artificial respiration was supplied, but without restoring the animal.

Post-Mortem.—The heart in diastole; great veins quite full; liver and kidneys considerably congested; lungs normal; stomach and intestines not congested; large veins rather full; arteries empty.

CXXXII

March 15, 1897.—Dog; weight, twenty kilos. Time of experiment, two hours and twenty-five minutes. *First.* As a control, incision in scrotum; slight fall in blood-pressure; respiration not changed. *Second.* Also as a control; manipulation of intestines; fall in central and portal, rise in the peripheral venous. Pressure quickly returned to its former level; respirations were slightly more shallow during manipulation. *Third.* Injected ten minims of a four per cent. solution of cocaine into the jugular vein, causing a slight rise in central, and a fall in the portal and the peripheral pressures. In two minutes the injection of an equal amount of cocaine produced similar phenomena. *Fourth.* Manipulation of testes after cocaine,—slight fall, then return to the former level. *Fifth.* Application of Bunsen's flame to the right hind-foot was now followed by a slight rise in the central, portal, and peripheral. *Sixth.* Bunsen's flame to nose caused general rise in blood-pressure; respiration unchanged.

Seventh. Passing glass canula into urethra ; no effect observed. *Eighth.* Intravenous injection of fifteen minims of a four per cent. solution of cocaine ; blood-pressure fell, heart-action became rather irregular, and respirations failed repeatedly. After a few moments the heart-action improved and respirations became better. *Ninth.* Several sharp blows were struck upon the upper surface of the head. Some of the blows caused fractures, but no effect was observed, excepting slight irregularities in the respiration. After two minutes had elapsed the animal was shot through the thorax with a .32-caliber revolver. Heart-action practically unchanged ; the blood-pressure began to decline greatly ; respirations not much affected, perhaps a little deeper and irregular. A second shot caused the respiration to become very shallow, then fail. Blood-pressure was quite low ; heart-action gradually grew weaker, and finally ceased.

Post-Mortem.—Heart in diastole ; bullet wound in pericardium and the right border of the heart ; a small clot in the cavity of the pericardium ; great veins quite empty ; abdominal viscera very pale ; both lungs lacerated by the bullets ; the pleural cavity contained a large quantity of clotted blood. The first shot passed in at the level of the third interspace, just at the left of the sternum, passing downward and towards the right, cutting through just posterior to the great veins, passing through both lobes of the right lung, through the diaphragm, the liver, and the right loin. The second shot entered the fourth interspace, just to the left of the sternum, passed through the inner margin of the upper lobe of the left lung, pericardium, right border of the heart, through the diaphragm and liver, wounding the hepatic vein, and passed out the right loin.

CXXXIII

Fox-terrier ; weight, fifteen kilos. Duration of experiment, two and a half hours. Chloroform and ether anæsthesia. Central pressure in right common carotid, peripheral in the left femoral. In adjusting the canula solution of magnesium sulphate was accidentally admitted into carotid. Convulsions followed, with lowering of pressure and cessation of respiration. Artificial respiration was practised for about thirty minutes. Applied Bunsen's flame to the paw ; respiration was immediately restored. After it was supposed that normal respiration would not again appear, and opportunities had been given for its restoration, Bunsen's flame to the right paw caused marked rise in pressure and establishment of respiration. The anterior crural and the sciatic nerves were injected with a four per cent. solution of cocaine, then the flame was applied to the foot as before ; fall in blood-pressure followed. In the control experiments, as well as in this, the dog was not under full anæsthesia. In the former the animal struggled on application of the flame ; after the injection of cocaine he did not. There was apparently blocking of the sensory impulses from the paw. Like experiments were made on the opposite paw, first as a control, and then by injecting cocaine into the sheaths of the sciatic and the anterior crural nerves. A circular skin incision was made around the thigh, so as to prevent possible impulses passing through the skin. The results in this case bore out those noted in the first experiment. The animal was finally killed by allowing the saline solution from the pressure-bottle to flow into the carotid. There was a straightening out of the limbs and a convulsive action, then death.

Post-Mortem.—Intestines very pale; liver and large veins full; left ventricle full, right empty; auricles empty.

CXXXIV

March 19, 1897.—Bitch; weight, thirteen kilos. Time of experiment, two hours and five minutes. *First.* Incised skin over left tibia, left surface; fall in blood-pressure. *Second.* Scraping off the periosteum, slight rise in blood-pressure. *Third.* Scraping periosteum, no change. *Fourth.* Sawed through tibia; the steady rise which was going on at time sawing began was unchanged. *Fifth.* Dilated the stomach with air; there was a fall in the central and a rise in the peripheral; respirations increased in frequency and in amplitude. *Sixth.* A blow over the stomach produced a fall in the central; respirations stopped temporarily. *Seventh.* A blow over the stomach, cardiac orifice, produced a fall in the central with shallow respiration. *Eighth.* Blow over stomach produced a decided fall in blood-pressure. *Ninth.* Blow over heart produced fall in the central blood-pressure, with small, short respiratory curves. *Tenth.* Slight blow upon neck, fall in central blood-pressure. *Eleventh.* Slight blow on "jugular;" no perceptible change. *Twelfth.* Blow on jugular, "knock-out" blow; rapid and decided fall in blood-pressure; respiration ceased; heart ceased temporarily, and resumed with "vagal" beats. Shot with .22-caliber ball. Ball entered the anterior surface of the neck above the larynx through the base of the tongue, through the left cerebral mass. Respirations stopped instantly. Heart beat on, then failed.

CXXXV

March 20, 1897.—Bitch; weight, nine kilos. Tracing taken while the canulæ were being inserted. Gradual rise in central pressure. Incising the skin over abdomen; slight rise. Incising uterus and removal of fœtus; slight fall in pressure, peripheral pressure corresponding to central. Manipulating the larynx; same as former results. Shot fired from a .32-caliber revolver, muzzle about six inches from the animal's head. The ball entered the superior surface of the head about one inch posterior to the superior margin of the orbit, and just to the left of the median line. The general course of the ball was down, slightly backward, and to the left. The fracture to the skull caused by the shot was very extensive. Upon tracing the course of the ball through the brain it was found that the left cerebral lobe had suffered great injury. The left lateral ventricle was injured and the choroid plexus, which caused a hemorrhage filling the ventricular cavities.

CXXXVI

March 21, 1897.—Dog; weight, twelve kilos. Blood-pressure was low. Three hundred and seventy-seven cubic centimetres of saline solution injected. Forcibly dragging upon the tongue caused slowing of respiration and slight inhibition of the heart. Cutting, puncturing, and crushing of the tongue caused no effect. Passing the finger into the mouth and forcibly thrusting it into the posterior nares caused partial inhibition of respiration and the heart. Touching the upper surface of the epiglottis, the soft palate, and the upper portions of the pharynx, no effect. Touching the under surface of the epiglottis near its base and touching the superior opening of the larynx caused considerable inhibition of respiration. Sawing through the lower jaw, no effect. Dragging the fragments of the lower jaw outward from each other caused a decided fall in the blood-pressure and slowing of the respiration. This was done several times with like results. Then dragging the halves inwardly past each other, applying considerable force, no effect was noted. Finally, opening the lower jaw to its normal extent, then forcibly pressing it open wider, caused slight fall in blood-pressure, slight slowing of respiration; in one instance turning the angle of the jaw very high caused an acceleration of the heart-beats. In no instance were the articulating surfaces of the jaws displaced. Dog was killed by firing a shot through the chest with a .32-caliber Smith & Wesson revolver. Respirations and heart both proceeded, and dog died of hemorrhage.

CXXXVII

March 22, 1897.—Dog; weight, ten kilos. Died of respiratory failure before the experiment began. Made some observations on injecting cocaine into the sciatic nerve and applied faradic stimulation above the block. A control had been secured before the cocaine was injected. Stimulation in the control caused sharp muscular contractions. Similar stimulations after the cocaine block caused but faint muscular contractions. Similar experiments were performed on opposite side with similar results.

CXXXVIII

March 27, 1897.—Dog; weight, ten kilos. Central pressure in femoral artery, peripheral the same. Cutting away the soft parts of the skull

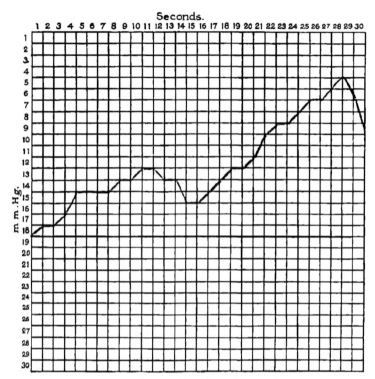

Seconds.

m m. Hg.

Cutting and crushing the skin.   Twenty-two experiments.

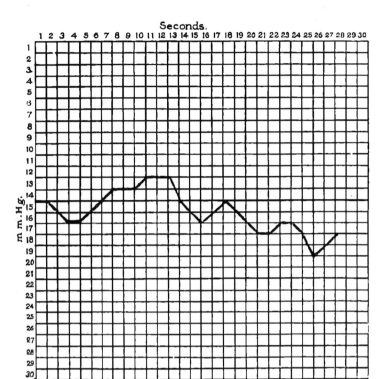

Seconds.

m.m.Hg.

Burning the scrotum. Ten experiments.

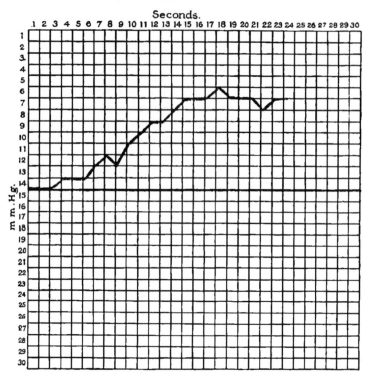

Seconds.

m m . Hg.

Burning the skin.   Fifty-eight experiments.

caused some fall in the central pressure. Considerable hemorrhage was encountered. Opening the skull with mallet and chisel, no change in respiration or circulation. Cutting away the skull with bone-forceps caused a slight rise in pressure. Extensively separating the dura mater caused a slight rise. Considerable hemorrhage was now encountered and a blood-clot rapidly formed under the dura mater, and as the tension increased respirations became more shallow, the heart-strokes became longer, and the blood-pressure rose very considerably. Finally, when the heart began to execute long vagal beats, the dura mater was incised, the pressure released, causing a fall in blood-pressure and restoration of respiratory rhythm. Tearing out the cerebral hemisphere arrested respiration; the central peripheral rose. Dog was then pithed.

Experiments Nos. 138 to 148, both inclusive, were devoted to observing the effect upon intracranial pressure produced by introducing a small pipette into the lateral ventricle and allowing salt solution from the pressure-bottle to flow. The results in these cases corresponded with those published in the extensive research of Horsley and Spencer. They have not, therefore, been detailed.

SUMMARY OF EXPERIMENTAL EVIDENCE

Remarks.—In this summary we have expressed as accurately as possible the results obtained in this research, and have added cuts illustrating the same. In the interpretation of these cuts it should be borne in mind that a mercurial manometer was used, so that the changes recorded would be more than thirteen times as great if expressed in the terms of a fluid having the specific gravity of the blood. Only immediate results were recorded, and the experiments were made under such varied conditions that in many of the subjects the curve was necessarily short. In a general way it may be stated that in the organs and tissues, such as the intestines, the experiments on which produced a primary fall in the blood-pressure, shock was induced earlier than such tissues and organs, as the extremities which produced a primary rise in the blood-pressure. It was observed that late in an experiment, when an animal had already been reduced to a state of shock, an experiment then upon organs producing primary fall or producing primary rise did not exhibit their characteristic rise or fall as in an experiment on an animal not so reduced. These averages are made up of experiments performed in every degree of shock. That is to say, it is made up of the early more marked as well as the later minor effects. An accurate idea of the nature of the typical curves may be obtained by consulting the preceding illustrations in the protocols. In order to make these illustrations as nearly accurate as possible, the averages were carefully measured by one who did not understand the purpose and objects of the experiments, and who spent six weeks of continued labor in making from the graphic records the exact measurements and calculations expressed by these cuts. In offering them, therefore, we do so with the assurance that they represent the results accurately. Finally, we wish to repeat what has been frequently asserted by experimenters, that no matter how faithfully the results may be recorded, the reader cannot receive so accurate an impression from the several experiments as can the experimenter himself, who makes the observations in the first instance.

TISSUES

SKIN

Cutting and tearing caused in the greater number of instances a rise in blood-pressure, though sometimes no effect was observed. On blood-pressure, excepting over the testes, and in many cases over the abdomen, there was usually a fall in pressure; late in the experiment not infrequently there was a fall. The rise in pressure was begun after the lapse of a few heart-beats, and ascended rather abruptly until the highest point was reached, when, after making a rounded crest, it fell again to the level it was before, or below this level; occasionally it remained higher. On incising the skin over the testes there was frequently a considerable fall in the pressure, in keeping with the phenomena attending injury of this organ. Burning the skin, including that over the scrotum, caused uniformly a marked rise in the pressure. While the peripheral pressure usually paralleled the central, it sometimes exhibited a contradictory movement in point of effect, and not infrequently it differed from the central in point of the time of its movement, though the tendency of the movement may have been in the same direction as the central pressure. This was construed as indicating separate vasomotor effects,—i.e., separate from the effect of the heart's action. Respirations were usually slightly accelerated and arrhythmic, though sometimes not appreciably affected. Burning the skin caused increase in respiratory action, sometimes a hyperpnœa. Mechanic or thermal injury of the skin covering the paws caused much greater circulatory and respiratory changes than like injuries in other regions of the skin. All the observations tend to show that the more specialized and abundant the nerve-supply to a part the more will it contribute to the production of shock when subjected to injury.

In a series of cases in which preliminary excision of the stellate ganglia was made, no alteration in the blood-pressure on repeating the above experiments was observed.

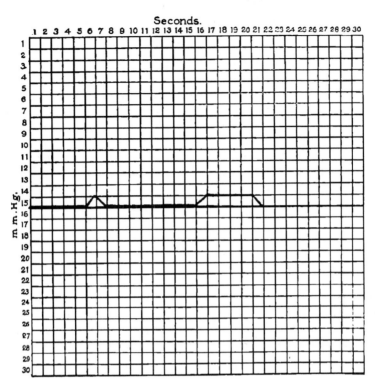

Seconds.

Cutting and crushing the skin; stellate ganglia excised. Eighteen experiments.

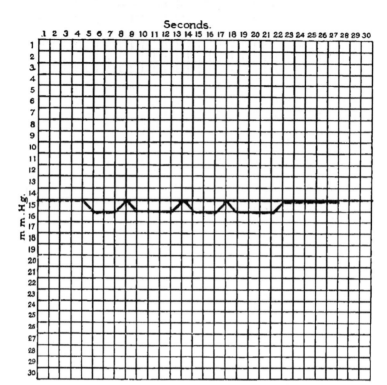

Cutting and crushing muscles.   Fourteen experiments.

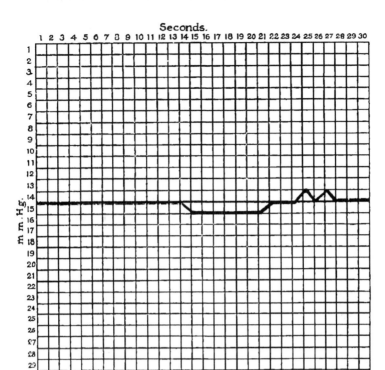

Crushing and sawing bone. Twenty experiments.

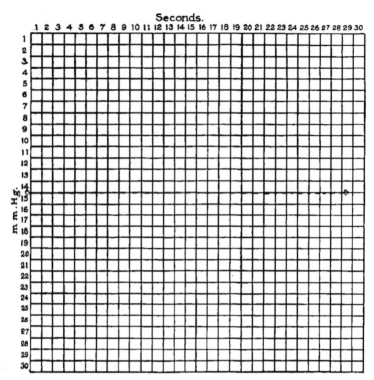

Seconds.

m. m. Hg.

Opening and operating on joints.   Ten experiments.

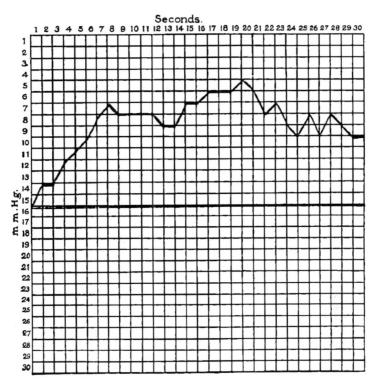

Stretching nerve-trunks,—e.g., sciatic. Twenty-nine experiments.

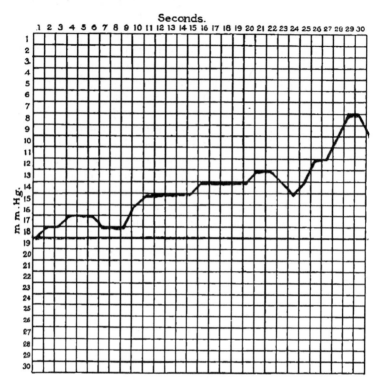

Crushing and cutting the turbinated bones, and forcibly dilating the nares.
Ten experiments.

CONNECTIVE TISSUE

No evidence was found of an appreciable effect from mechanic or thermal injury to connective tissue, fascia, tendons, ligaments, etc.

MUSCLES

Cutting or crushing muscles was attended by practically the same phenomena as described under the heading Skin, but to a much less degree. In many instances extensive cutting of muscles caused no appreciable change. As a rule, the simple skin incision in a hip-joint amputation had at least as much effect upon the circulation and the respiration as cutting the entire muscular mass of the thigh. No differences could be determined in the various skeletal muscles.

BONES

PERIOSTEUM.—In several instances roughly separating the periosteum from the bone caused a small rise in blood-pressure similar to that observed in injury of the skin and muscles; in other instances no effect was noted. Sawing through bone whose periosteum had not been previously removed sometimes was attended by slight rise in pressure and corresponding respiratory changes; but sawing, cutting, crushing, and breaking bones whose periosteum had previously been removed was not in any instance attended by any alteration in either the blood-pressure or the respiration. The same may be said of cartilage.

JOINTS

On cutting, sawing, curetting, or crushing the joints no appreciable effect was noted. The same may be said of opening the hip-joints and other large joints and doing the various surgical operations thereon.

NERVE-TRUNKS

ON CIRCULATION.—Crushing, tearing, contusing, or cutting with dull instruments causes usually more or less rise in pressure, followed by a fall occasionally to the former level, but usually lower. Sometimes there was a very great decline below its former level. Vasomotor curves frequently become more prominent just after the curve of stimulation. By repeating the irritation, the mean blood-pressure, losing somewhat after each experiment, suffered depression in proportion to the intensity and the duration of the irritation. In some instances, especially in observations after repeated irritation had been made, there was an immediate decline in pressure; the decline was usually quite gradual, and continued longer than the upward tendency in other instances. Severing the nerves quickly with sharp scissors usually gave comparatively little effect; if a rise, it was but momentary ; if a fall, it was immediate, and usually was not recovered from. The total disturbance was decidedly less in quickly severing with sharp instruments than in contusing, tearing, dragging, etc. Electric and thermal irritation both caused curves very similar to, though more marked, than mechanic irritation.

The peripheral, the arterial, and the venous pressures, in the femoral artery, femoral vein, carotid artery, and jugular vein, in the greater number of instances, partook of the general movements of the central pressure. In some instances there seemed to be independent action, similar to that described under the heading Skin. In cases in which the stellate ganglia had been removed, there was usually no change in any of the pressures on the application of the above-mentioned tests, neither was there usually any change in cases in which the cardiac branches had been previously severed. In several instances there was a slight rise, and in a few some fall, on the application of such stimuli. However, the contrast, even in the apparent exceptions, was striking,—*i.e.*, there was but little alteration in the circulation in the cases subjected to preliminary excision of the stellate ganglion. *After the animal had become very weak, or after repeated irritation of a given nerve-trunk, there was sometimes a fall without any preliminary rise.* In some cases in which cocaine or eucaine had been injected, and but little time had elapsed, say only a minute, there was a fall. The nerve-trunks referred to are the peripheral only.

On Respiration.—There was usually an alteration in rhythm ; irregular amplitude, short strokes and long strokes followed each other in disorder, as did also over-inspiratory and over-expiratory tonus ; but, on the whole, the total action is decidedly increased. Normal respiratory rhythm is usually promptly restored on cessation of stimulus. In cases in which there was subsequent effect, it was usually manifested by a slow rhythm, with either diminished or increased amplitude. This refers to nerve-trunks of the extremities only.

In all forty-three observations were made.

REGIONS

HEAD

Nose.—Mechanic injury of the nares, involving contusion, laceration, and dilatation, and in some instances crushing of the turbinated bones, caused a distinct rise in the central blood-pressure, the ascent of the curve being not so sharp as in similar injuries of the extremities. No separate vasomotor effects were observed. Respirations were slowed, and the amplitude diminished in most instances. An exception in the way of an increased depth was occasionally noted. Thrusting the finger into the

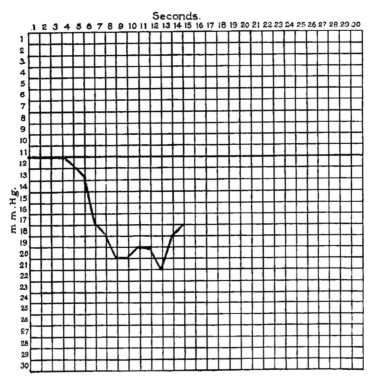

Severe manipulation of the posterior nares. Ten experiments.

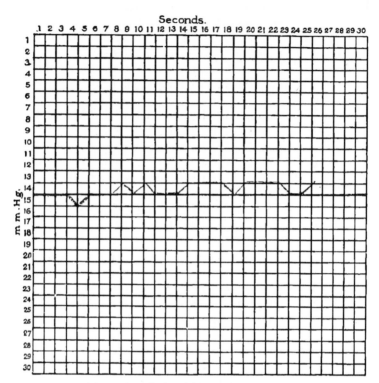

Injury and enucleation of the eye. Ten experiments.

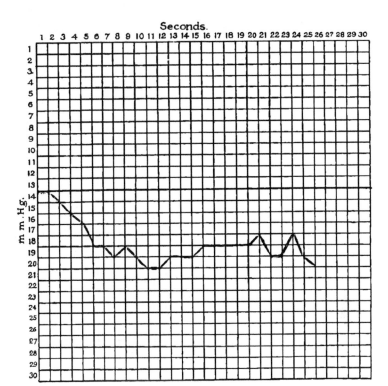

Injuries to the internal ear. Ten experiments.

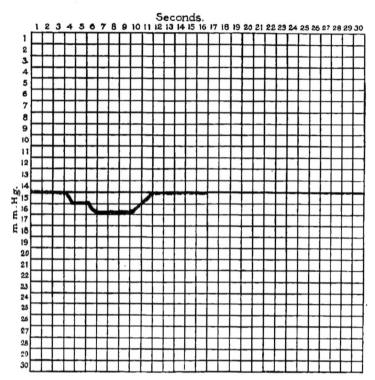

Dragging on the tongue. Ten experiments.

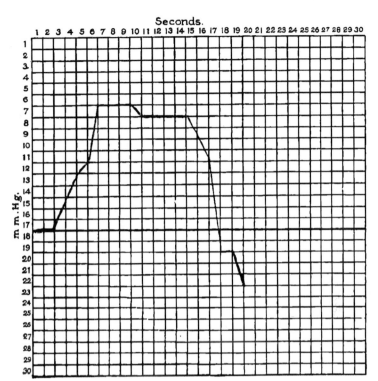

Forcibly separating the lateral halves of the severed inferior maxilla. Ten experiments.

posterior nares and imitating forcible removal of a tumor caused in some instances a partial inhibition of the respiration and of the heart.

EYES.—Mechanic injury of the conjunctiva, of the structures of the orbit, of the eyeball, and of the optic nerve occasionally caused a little, though usually no appreciable change in either the circulation or the respiration.

EARS.—The external ear, when any effect was noted, produced such alterations as were described under the subject of skin. On contusing, puncturing, and otherwise destroying the structure of the internal ear there was usually a fall in the central, portal, and cephalic pressures, and a rise in the peripheral venous; in one instance there was a fall in the portal only. In the other instances there was a rise in the blood-pressure, and in but one was there any effect on respiration,—a diminished frequency and depth.

MOUTH.—On crushing, cutting, tearing, and puncturing the tongue there was no effect on either the circulation or the respiration; on forcibly dragging the tongue out of the mouth there was partial inhibition of the respiration and the heart. No effects were observed on operations on the buccal cavity. Crushing the jaws by turning down the screw-clamp of the dog-holder produced no appreciable effect. After the jaws had been placed in the position of their maximum normal separation, by applying considerable force in further separating them respirations were considerably inhibited and the heart-strokes took on a slight "vagal" character,— i.e., the frequency was diminished and the length of stroke increased. On sawing through the median line of the lower jaw, then separating the severed sides laterally from each other, there was a very sharp and considerable rise in blood-pressure, with decided acceleration of the heart. This was repeated a number of times, sometimes moving them singly, sometimes moving both laterally outward, and it always markedly altered the blood-pressure and partially inhibited the respiration. In one instance, in an extreme outward and upward displacement, there was a decided cardio-inhibitory effect. In no instance was the condyle dislodged from its articulation. These effects were very probably due to stretching or other mechanic injury of the nerves governing the respective functions whose disturbance was observed.

PHARYNX.—Mechanic injury of the vault of the pharynx, the soft palate, the velum, tip and upper surface of epiglottis, and base of the tongue produced varying effects. The under surface near the base of the epiglottis and the superior laryngeal opening caused respiratory arrest on even very slight irritation. On more vigorous digital manipulation there were cardio-inhibitory effects as well. As to the remainder of the pharynx, excitation of swallowing was noted in a comparatively small zone of the lower pharynx, extending just above and below the level of the superior laryngeal opening. Dragging on the soft palate slowed the respiration, and in some instances in which very considerable roughness of manipulation was practised on the vault and soft palate there was very marked slowing of respiration and of the heart-beats.

BRAIN.—*Dura Mater.*—Gentle separation of the dura mater from the skull was attended by but little effect. *Hemispheres.*—Cutting out the cerebral hemisphere caused a marked fall of the central and a rise in the peripheral arterial blood-pressure. There was a very striking irregularity in both the blood-pressure and the respiration. These experiments were attended by a considerable loss of blood. Respirations failed during the removal of the hemispheres. Gunshot wounds of the hemispheres caused sometimes enormous sweeping " vagal" heart-beats, sometimes an acceleration, usually the former. Respiration in every corresponding case was instantly arrested and rarely resumed. In two instances, after the lapse of some time very slow and shallow respiratory efforts were noted. Ten observations were made on the effect of rapidly applied pressure within the brain, as might occur in sudden hemorrhage. In each case the heart-strokes took on a " vagal" character and the respirations failed first. The method employed consisted in making a very small opening into the skull, through which a slender glass canula connected with a pressure-bottle of normal saline solution was connected. The canula was passed into the lateral ventricle and the solution allowed to flow. The respiration rapidly failed in each instance, and the heart continued beating, exhibiting very striking " vagal" strokes. The production of depressed fracture by a hammer-blow caused instant failure of respiration; the heart beat on strongly for a short time, then failed. Blows not severe enough to cause fracture, but a considerable jarring, caused occasionally a sweeping heart-beat and a temporary arrest of respiration. Pressure upon the surface of the brain produced like results. Mallet-blows if rather severe produced marked disturbance in respiration and in the cardiac action. Injury of the medulla caused a momentary rise, followed by a staggering fall of the blood-pressure to zero.

NECK

LARYNX, TRACHEA, AND ŒSOPHAGUS

LARYNX.—Contact, even slight, with the mucous membrane of that portion of the larynx extending from just below the vocal cords to the outer margins of the superior laryngeal opening, including the under surface of the epiglottis, caused instant respiratory arrest under surgical anæsthesia. In cases in which contact was prolonged the respirations gradually resumed their normal rhythm. In cases in which contact was made more firm, as in dilatation of the larynx or compression between

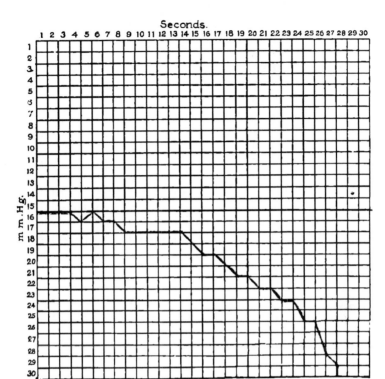

Cutting away the cerebral hemispheres. Ten experiments.

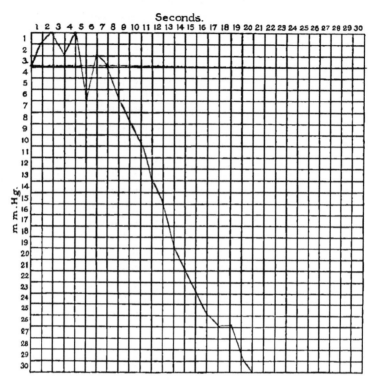

Seconds.

m.m.Hg.

Crushing the medulla. Ten experiments.

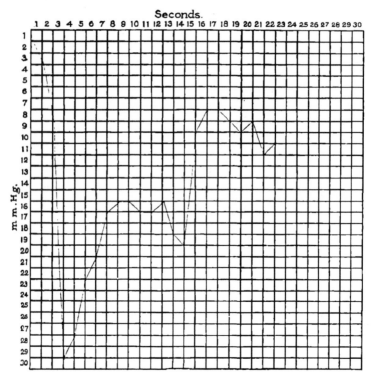

Seconds.

Manipulation of the larynx. Ten experiments.

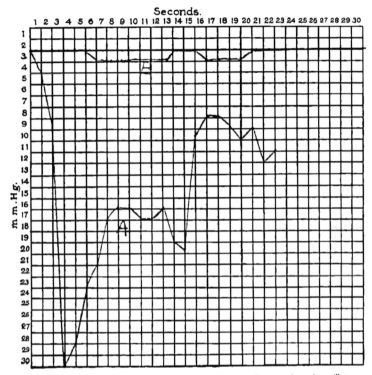

Seconds.

m.m.Hg.

A, manipulation of larynx; *B*, manipulation of larynx after use of cocaine. Ten
experiments.

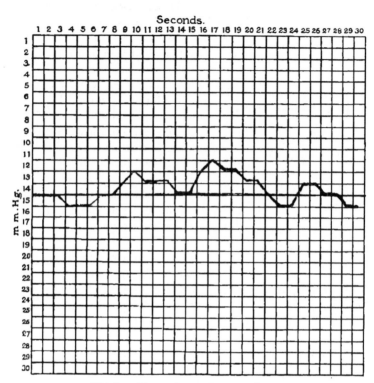

Seconds.

Dilatation of the œsophagus.   Seven experiments.

extra- and intra-laryngeal force, in addition to the respiratory arrest there was partial or complete inhibition of the heart. On continuing intra-laryngeal contact tolerance was acquired. Severing the inferior laryngeal nerves did not prevent the appearance of the above phenomena; severing the superior laryngeals abolished them; hypodermic administration of atropine prevented the cardio-inhibitory effect of such manipulation. Cocaine thus administered had a like effect. Neither drug, however, so employed, prevented to any appreciable degree the respiratory inhibition. Cocaine locally applied, even in one-half per cent. solution, upon the laryngeal mucosa, within thirty seconds entirely prevented the respiratory and cardiac inhibitions on repeating a like manipulation. The larynx might then be cut, stretched, contused, and otherwise injured without any manifest inhibitory effect on either the respiration or the heart-action. In various injuries and operations upon that portion of the larynx inferior to the *inhibition area* above described and all the accessible portions of the trachea, no effects were noted. The same may be said of the œsophagus, excepting in dilatation of the latter, which had a variable effect,—in one case sharp rise in pressure and irregular, shallow respiration. In five other tests there was observed a temporary fall in blood-pressure, with diminished pulse-volume, probably mechanically produced; the respiratory movements were increased in amplitude. Performing operations, such as œsophagotomy, and inflicting injuries, as stab and gunshot wounds, produced no observable effects on either the circulation or the respiration.

OTHER TISSUES OF THE NECK.—When proper respect was paid to the vagi and to the sympathetic nerves, extensive dissection was made without disturbing either the circulation or the respiration, but mass-traction on the wound, or forcible blunt dissection of the deeper tissues, usually caused cardiac or respiratory inhibition, or both. This is especially true of the upper portion of the neck on a level with the larynx and above it, including the region below and behind the angle of the jaw. The so-called "blow upon the jugular" in sporting parlance owes its effect to the mechanic stimulation of the vagus, thus inhibiting the heart. After preliminary administration of atropine or section of the vagi distal to the point of injury no inhibitory effects were produced. In two instances, however, a rise in blood-pressure with an acceleration of the beat was observed; in other observations the typical long, sweeping "vagal" beats appeared. The blows in the former instances probably missed the vagus and stimulated the sympathetic. Gunshot wounds of the neck have no effect aside from a momentary one, excepting as they may injure the vagus, the sympathetic, or the "inhibition area" in the larynx. Firing a shot through the thyroid cartilage has but a transitory inhibitory effect.

Through the "arrest" area there was temporary inhibition of respiration and "vagal" strokes of the heart. Grasping the neck, as in powerful choking or strangling, caused inhibition, partial or complete, through irritation of the "inhibitory" area of the larynx or traction on the superior laryngeal nerves. It was quite difficult to produce sufficient mechanic injury of the vagi by grasping the intact neck to cause any inhibition.

THORAX

In resection of ribs, fractures, stab wounds, gunshot wounds, etc., in experiments in which the thoracic cavity was not opened, there was merely the effect of such injuries of skin, muscle, and bone. Injury of the skin in this region did not produce so profound changes as in that of the foot and leg. Respiration, however, was considerably affected in rhythm in tearing or cutting the extraordinary muscles of respiration. The tearing and the dragging effects were probably due to a corresponding injury of the supplying nerves. Opening the thorax is attended by the greatest irregularity of breathing,—irregular rhythm and excessive action,—the blood-pressure undergoes sweeping changes in height, and the heart-strokes usually become short and very irregular. In no other manner were there produced such exceedingly irregular blood-pressure and respiratory tracings as in the intra-thoracic procedures. Artificial respirations were, of course, maintained, and naturally interfered with the normal efforts, and, besides, the mechanic factor in operating on the chest-wall was considerable. These considerations somewhat lessen the value of the respiratory tracings, but direct observation of the respiratory action made it more than probable that the heart-action, which was perfectly recorded, was even more disturbed than was the respiratory. In resection of the anterior chest-wall, by placing a double row of ligatures around the ribs and the sternum, then cutting between the ligatures with bone-forceps or saw, and exposing the entire chest cavity by elevating this flap, there was almost immediate normal breathing and heart-action after fastening down the resected wall by tying the corresponding long ends of the ligatures and suturing the skin. Blood-pressure usually suffered a considerable decline during the performance of this manœuvre. This decline was not due to hemorrhage, as the operation was practically bloodless. This operation was done ten times without a death, and, although there was a double pneumothorax, competent normal respiration was in each case restored on closure of the chest. The respiratory action was usually greatly accelerated for a considerable time afterwards. However, in experiments in which the chest had been opened there was a tendency to recurring respiratory failure in the subsequent course of the experiment.

HEART.—The slightest direct contact with the heart caused marked changes in its beat and in the blood-pressure. Touching or gently pressing upon the pericardium over the apex caused a staggering, immediate fall in blood-pressure, with short, irregular strokes. On removal of the contact blood-pressure immediately mounted again. Likewise, on touching the base and the large blood-vessels near the base, there was a staggering fall in the blood-pressure with irregular stroke, but a more sweeping stroke than in the apex experiments. Gently displacing the heart laterally caused extreme irregularity, the heart executing a series of most irregular sweeps, with great fall in blood-pressure. This fall was recovered from upon the release of the pressure. In like manner, pressing up the diaphragm with the hand so as to feel the heart-impulse caused phenomena similar to the above, but to a less marked degree. Picking and holding up the pericardium, preparatory to incising it, caused a rise in pressure followed by a decline; when released there was a very great rise. Incising the pericardium produced but little effect. When the heart had been much weakened, touching or gently moving it caused sometimes complete cessation for a time, then slowly it resumed its beat. Puncturing the heart with a scalpel caused only a lapse of one to several beats. A gunshot wound of the heart, not penetrating the chambers, caused but temporary arrhythmia for several beats.

LUNGS.—Mechanic injury of the lungs in the way of manipulation, contusion, stab wounds, gunshot wounds, etc., on the whole seemed to affect the heart more than the respiration. There was great difficulty in making satisfactory observations on this point. Some observations showed very marked " vagal" heart-beats on pinching the lungs with the fingers.

LARGER VESSELS.—The venous trunks especially cause sweeping changes in blood-pressure by mechanically interfering with the flow of blood into the chambers of the heart, and also in manipulation near the base of the heart the latter's rhythmic contraction is interfered with; probably because the contraction-wave at the commencement of each cardiac cycle begins in the venous trunks near the base. In fresh animals there was a prompt recovery of a lost cardiac equilibrium, but not so if the animal had been well exhausted when the equilibrium was disturbed.

DIAPHRAGM.—Contact, however slight, with the abdominal side of the diaphragm caused in every instance markedly arrhythmic respiration. If contact was extensive, every part of the respiratory curve became irregular and fragmentary. Puncture of the diaphragm and gunshot wounds usually caused immediate arrest of respiration, and if the arrest proved to be but temporary, there was a tendency to respiratory failure later in the experiment. Hot water applied to the abdominal side of the diaphragm very greatly augmented the amplitude of the respiratory movements and increased its frequency.

ABDOMEN

INCISION THROUGH THE ABDOMINAL WALL.—In making the incision through the skin in the abdominal sections there was frequently noted a fall in the blood-pressure; this, in fact, was the rule; making control experiments by producing slight pressure upon the abdominal wall, a rise was observed, but the rise was only temporary. Cutting muscles and fascia produced little or no effect. On opening the peritoneum a fall was noted; in some cases there was no immediate effect. On exposing the abdominal contents *in situ* to contact with air, there was usually a gradual fall after some time had elapsed.

PERITONEUM.—Contact, however slight, with the parietal or the visceral peritoneum caused markedly arrhythmic respiratory action, the amplitude was markedly, sometimes very greatly, diminished or increased, and the curve broken and irregular. The diaphragmatic peritoneum produced the most marked respiratory changes. Continuation of the manipulation does not secure tolerance unless confined to the same area. We have many evidences of exhaustion of the respiratory apparatus by continued peritoneal excitation. The application of hot or cold water, especially the former, caused a very great increase in rapidity and depth of respiration. Exposure of the peritoneum or its manipulation caused rapid dilatation of the vessels of the mesentery and of the hollow viscera. The viscera became at first red, and after further exposure gradually livid. The rapidity of the development of the lividity was somewhat in relation to the respiratory as well as the circulatory disturbance. The mesenteric veins became more prominent, especially the small venous radicles at the base of the intestines. On long exposure and great irritation by manipulation, even the clear, transparent peritoneal spaces in the mesentery displayed vessels and sometimes became red. The arteries at first seemed larger and pulsated more distinctly, but later, when the blood-pressure had become quite low and the intestines livid, scarcely any pulsation was visible. With the development of these vascular changes the blood-pressure *pari passu* declined. The more severe the injury, the greater the extent of contact and exposure, the more quick and rapid was the decline of the blood-pressure. In a few instances there was a preliminary rise, giving way to a decline later. Then, again, in occasional instances, there was no notable change in blood-pressure until as much as half an hour had elapsed. Extensive and continued manipulation for as long as twenty minutes, and in one case half an hour of experiment, scarcely

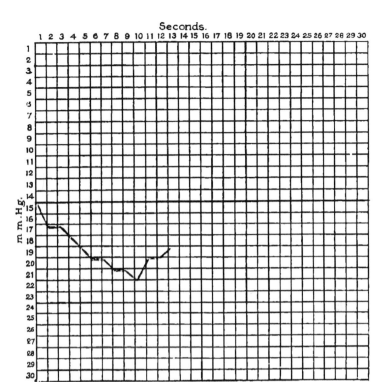

Incising skin of the abdomen. Ten experiments.

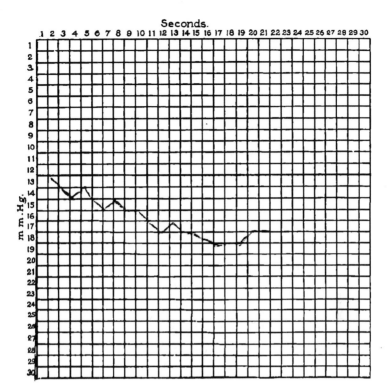

Manipulation of the parietal peritoneum. Ten experiments.

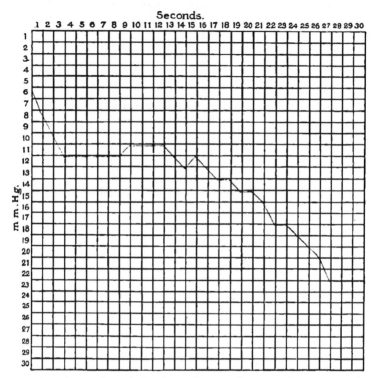

Seconds.

m.m.Hg.

Exposing and manipulating the intestines. Fifteen experiments. Each square represents three seconds.

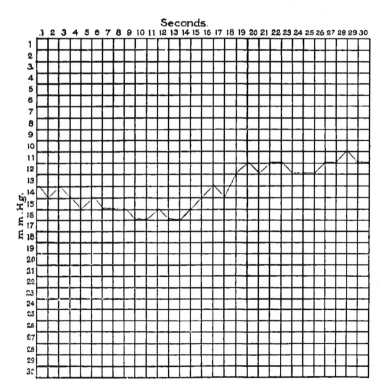

Manipulation of intestines with the splanchnic arteries clamped. Eleven experiments.

altered the pressure. In weakly dogs, or later in an experiment, the rate of decline in pressure seemed proportionate to the animal's condition. A water-manometer in the splenic vein, with its canula pointing towards the heart, showed a decided rise in a number of observations during the developments of splanchnic shock; while the central blood-pressure was declining the central portal pressure was rising, showing an increase of blood in that area. In a series of experiments the arterial supply of the splanchnic area, that is the cœliac axis, and the mesenteric arteries were clamped subperitoneally by making an incision along the anterior border of the rectus muscle and separating the peritoneum up to the diaphragm and over the abdominal aorta. With the splanchnic arteries thus clamped, no amount of manipulation caused the characteristic decline as before. Directly upon withdrawal of the intestines, there was in some instances a sudden fall in the central pressure, with immediate compensation following. This sudden fall may have been due to mechanic interference with the inferior vena cava or its tributaries. On slapping the intestines it was noted that they contracted, became paler, and the central pressure would rise instead of fall. These observations were made a number of times. Attempting to find an explanation of this additional rise, the cardiac branches of the stellate ganglia were severed in addition to clamping the splanchnic arteries. On repeating like treatment of the intestines there was sometimes observed a slight rise in the central blood-pressure; but, apparently, not nearly so marked as before. The rise which now occurred could not be repeated, while formerly it could be repeated on repeating the slapping and striking of the intestines. It would appear, then, that there are also acceleratory impulses which pass from the intestines through the cardiac branches of the stellate ganglia to the heart, and possibly to the vasomotor centres as well. The slight rise, which could not be made to repeat itself after the cardiac branches had been cut out, was probably due to muscular contraction of the intestinal walls and contraction of the veins, causing an additional amount of blood to flow to the heart, and the veins not receiving a new supply, an additional amount of blood sufficient to cause this rise in central pressure could not again be forced out of the comparatively empty veins by such stimulation. Clamping the superior mesentery alone very greatly diminished the usual rapidity of the development of splanchnic shock. With all the splanchnics clamped, if in addition the abdominal aorta was clamped, there was, as usual, a considerable immediate rise in the central pressure. However, this rise was maintained, and the usual prompt compensatory fall without clamped splanchnics did not occur. Such observations were repeatedly made, and support the belief that the splanchnic area is the important compensating or regulating area of the circulatory apparatus. When the splanchnics were unclamped there was a staggering fall in the central blood-pressure, and although a compensatory rise was promptly inaugurated, it in no instance reached the mean-pressure level. When compensation had developed as far as possible and a further intestinal manipulation was made, an immediate further fall was noted. While an animal was quite fresh, compensation, after a fall in pressure incident to the dilatation of the splanchnic vessels, occasionally did occur; but it was usually not complete. In a weak animal, or late in an experiment, compensation was rarely observed even to the slightest degree. The decline was steady until death. The blood-vessels of the large intestines are comparatively less affected than the small, the gastric vessels more than those of the large intestines, but probably less than those of the small intestines. The pelvic peritoneum, as nearly as comparative observations permit deductions, contributed less to the production of shock than the abdominal peritoneum proper. The omentum is precisely the antithesis of the foregoing. When any effect was obtained, and usually no effect was observed, it was a rise of blood-pressure, a rising curve quite comparable to that following injury of the skin. The omental vessels manifested but slight dilatation after severe and continued irritation, during which the vessels of the hollow viscera were extremely dilated. Extra-peritoneal dissection did not affect either the respiration or the blood-pressure notably, unless the splanchnic nerves had been involved, when there followed a decided dilatation of the splanchnic vessels and corresponding fall in blood-pressure.

LIVER.—Aside from the mechanic effects from forcing blood from this vascular organ by pressure and the effects of contact with its peritoneal covering, as above indicated, no special effects were observed. Dilatation of the cystic duct caused no appreciable effect. Manipulation of the gall-bladder caused a marked temporary fall; but it is extremely probable that this was due to mechanic interference with the blood-current in the larger venous trunks lying in such close anatomic relation.

KIDNEYS.—Cutting, contusing, crushing, or any other mechanic injury of the kidney, caused no notable effect in a single instance, except when there chanced to be peritoneal contact. Later there was usually some decline, due probably to irritation of the peritoneum and the splanchnic nerves. The same may be said of nephrectomy performed a number of times.

SUPRA-RENAL BODIES.—On similar treatment a rise in the blood-pressure was noted in a number of instances. No separate vasomotor effects were noted.

SPLEEN.—No special results were noted other than when that organ was compressed. There was an immediate small rise, probably due to forcing out blood from its vascular meshes. Splenectomy was performed a number of times and with practically no effect.

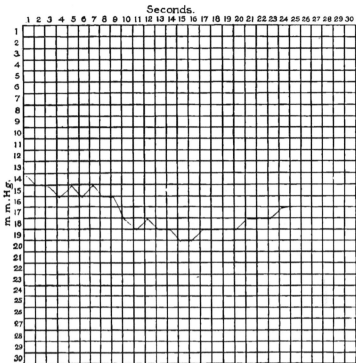

Manipulation of the kidneys.   Eleven experiments.

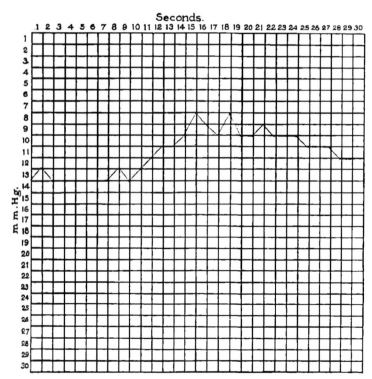

Seconds.

m.m.Hg.

Manipulation of the supra-renal bodies.   Ten experiments.

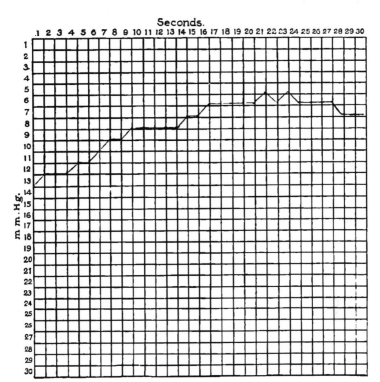

Seconds.

m.m.Hg.

Manipulation of the uterus. Fourteen experiments.

Incising scrotum. Ten experiments.

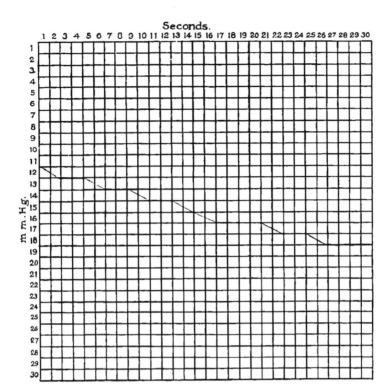

Manipulating and crushing the testicle. Forty-one experiments. Each square represents one-half second. Many of the experiments exhibited a sudden and profound fall, not represented in the average.

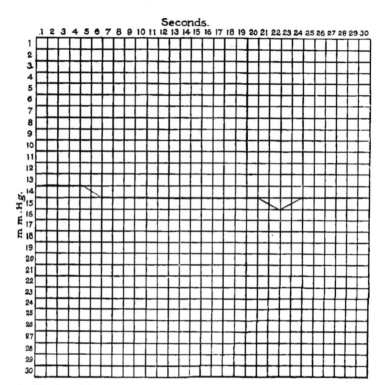

Crushing the testicle after injecting cocaine into the cord. Ten experiments. Each square represents one-half second.

BLADDER.—Cutting, compressing, over-distending, or otherwise injuring the bladder, caused a rise in the blood-pressure, if any effect at all was produced. In many of the observations no effect was noted. Pressure upon the full bladder produced the most marked results.

UTERUS.—Incision, contusion, manipulation, or any other mechanic injury of the uterus, caused uniformly a rise in the blood-pressure. The rise appeared rather slowly, but was in many instances very marked. Sometimes the pressure gradually declined to its former level, but tended to remain, for some time at least, at the level to which it rose. Repeating the injury at intervals before compensation occurred raised the mean pressure very considerably, comparable to the rise caused by clamping the abdominal aorta. Cutting the cardiac branches of the stellate ganglia does not prevent the rise. While conclusive proofs are not at hand, there is considerable evidence that the rise is due to vasomotor action. What has been said of the uterus may be said of the ovaries and oviducts and in importance with the order given. Ten bitches each furnished repeated observations on these points. Cæsarian section was twice made, and noted for the absence of any changes in the pressures during operation.

MALE GENITAL ORGANS

TESTICLES.—Cutting the testicles, spermatic cord, *tunica vaginalis*, and frequently even the skin of the scrotum, caused, in most instances, a fall in blood-pressure, appearing after a short interval. Manipulation, though gentle, of the testicle nearly always caused a very marked fall in the central blood-pressure. The same observations were made with regard to the spermatic cords. In short, any manipulation of any part of the testicle—its coverings or the spermatic cord—was attended by a marked, sometimes exceedingly great, fall in blood-pressure. While the central pressure was falling, the portal was usually as markedly rising, as evidence of the splanchnic dilatation as the cause of the fall. The blood-pressure usually recovered completely, or nearly so, its former level. These observations were made fifty-three times, and occasionally, though rarely, no fall was marked. Respirations were slowed and shortened, and in some instances irregular. Injecting cocaine into the organ, into the *tunica vaginalis*, or into the spermatic cord after a control had been first obtained, then repeating like operative procedures, no fall occurred in central and no rise in portal pressure; neither were the respirations altered. Injecting a sufficient quantity of atropine into the jugular vein after securing an inhibitory laryngeal control and proving the atropine competent to abolish cardio-inhibitory impulses, the testicle was then subjected to experiments similar to the above, which was followed by the usual fall in pressure; neither was the fall prevented by preliminary section of the vagi. This is taken as evidence that the fall is probably not due to a cardio-inhibitory effect. Neither did jugular injections of cocaine prevent the fall in the pressure, though, as nearly as could be judged, the fall was considerably less than in the controls.

PENIS.—The same may be said of the penis, though the alterations were not observed to be produced in nearly so marked degree.

VAGINA.—Forcibly dilating or otherwise injuring the vagina caused an increase in depth and frequency of the respirations, and usually a rise, though occasionally a decline, in blood-pressure.

ANUS.—Forcibly stretching the rectum and anus caused usually a rise, sometimes a fall, in blood-pressure and an increase in the frequency and the depth of respiration. There is considerable evidence that the fall in pressure was due to mechanic interference with the flow of blood in the venous trunks,—since the fall appeared almost instantly, and the manometer in the femoral vein showed a rise at the same time that the central pressure was falling.

EXTREMITIES.—Cutting, crushing, fracturing, amputating, and burning the extremities were usually attended by a preliminary rise in blood-pressure, followed by a fall usually lower than before the operation. Respirations were much altered in rhythm and temporarily increased in frequency. There is no evidence of the correctness of the opinion expressed by many clinicians that there is considerable shock produced in sawing through the bone in amputations, nor of similar opinions as to opening large joints. The error probably arose in confusing the effects of cutting, traction, or other injury of the nerve-trunks, which would be clinically manifested about the time a rapid operator would have proceeded to sawing the bone or opening the joint in a hip-joint amputation. Traction on the nerve-trunks especially caused marked effects upon both respiration and circulation. The shock is in direct proportion to the amount of excitable tissue injured and the time occupied in the operation. The experiments on which these observations are based were made without loss of blood. Injury of the paws is more productive of shock than that of any other superficial part of the extremities. There is sufficient evidence to show that the shock produced is in direct proportion to the nerve-supply and to the functional importance of the part injured. In operations involving the brachial plexus the respiratory changes were very striking, the frequency of respiration was much increased, and over-anæsthesia was accordingly produced unless the anæsthetic was removed. Operations involving the brachial plexus produced also very marked vasomotor disturbance, as shown by the peripheral manometer and by the Traube curves in the central blood-pressure tracing.

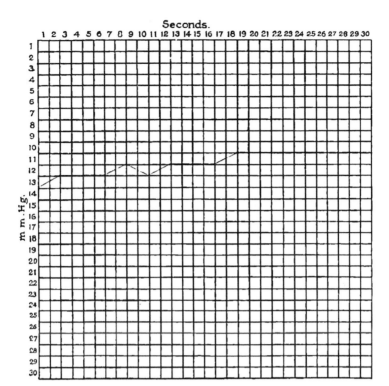

Seconds.

m.m.Hg.

Forcible dilatation of the vagina. Ten experiments. Each square represents one-half second.

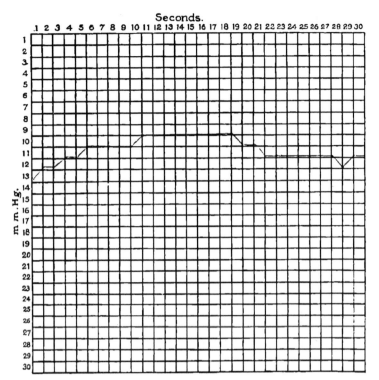

Manipulation of the ovaries. Six experiments.

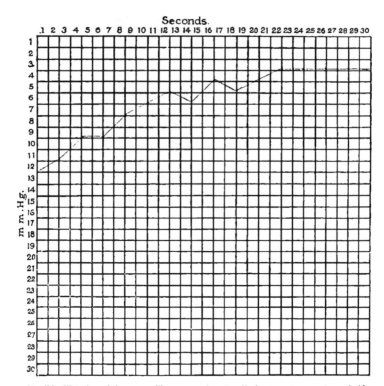

Forcible dilatation of the anus. Eleven experiments. Each square represents one-half second.

FACTORS CAUSING SHOCK

DURATION OF OPERATION

The duration of an operation was found to be an important factor in the production of shock. Animals may be killed by the effect of continuous anæsthesia alone, though the anæsthetic is carefully administered, so that a percentage, calculated upon the ratio between the actual duration of anæsthesia and the average length of time a dog may survive continuous anæsthesia, is allowed the pure anæsthetic factor in any given case; that is to say, if ten hours be allowed as the average length of time a dog may live under continuous anæsthesia, and the given experiment lasted two hours, then twenty per cent. of the cause of death was calculated to represent the anæsthetic factor. This calculation applies to ether. There is strong evidence tending to show that chloroform, even barring accidents, is a more potent factor in destroying the animal than is ether. The element of time in relation to the exposed field may be stated as follows: contact with air is a very great irritant to local tissues, owing to the lowering of local temperature and to the drying. The effect of exposure is strikingly observed in case of the peritoneum, pleura, and brain. Exposure affects particularly the vasomotor mechanism. If a bloodless field of operation, the thigh for example, be exposed, it soon becomes suffused with blood, all the vessels become dilated, the translucency of the tissue is lost, and further dissection then becomes bloody. The brain, when exposed, soon becomes abnormally irritable, and, in faradizing the cortex over a given motor area, say the fore leg, there will likely be an overaction. Many adjacent centres may become excited, and not infrequently a general convulsion is produced. However, after a long time, the hyper-excitability is followed by non-excitability. In the fresh state, or when the brain is kept covered with normal salt solution, within certain limits of time, the motor discharges correspond with the area of cortical stimulation. The same may be said of nerve-tissues generally, and the importance of preventing exposure of the vagi, the splanchnic, and other special nerve-tissues must be ever in mind. Exposure of the thoracic cavity causes great disturbance of respiration, and the time of exposure should be as short as possible. Exposure of the capacious splanchnic area is attended by a rapid dilatation of the splanchnic vessels, leading to intense congestion, detracting thereby a dangerous amount of blood from the somatic circulation, and inducing a rapidly declining blood-pressure. Even the apparently bloodless spaces in the omentum on exposure become red. Not only does exposure of the peritoneum cause splanchnic vascular dilatation, but the respirations are unfavorably affected as well. The element of time in abdominal operations in every experiment was unmistakable.

In animals subjected to extensive removal of integument, shock was induced with a rapidity proportional to the area of exposure, and its depth corresponded with the duration of the exposure. The exposure of the tissue, as a factor in the production of shock, was believed to bear a direct ratio to the area and the duration of the exposure.

TEMPERATURE

Observations on this point must necessarily be relative, but it seemed to us that when the opened tissues were exposed to cold air in the laboratory, during winter experiments, when the heating apparatus had been temporarily out of order on account of freezing, the animals seemed to succumb more readily than under the ordinary warmth. The effect on the intestines of cold water and of the intravenous cold saline solution showed more directly the depressing influences of the cold. Then, again, the direct effect of warm towels applied to the exposed intestines, of warm saline in the abdomen, improved the respiration immediately, and, as nearly as could be estimated, caused at least a check in the declining blood-pressure. The striking contrast between the effect of cold and warm salines on the exposed nerve-fibres, and especially on the exposed cortex cerebri, was repeatedly noted. In electric stimulation of the motor areas it is entirely essential to keep the cortex warm in order to obtain the best motor stimulation; indeed, a cool cortex very soon becomes non-excitable. In a number of observations on the rectal temperature, a gradual decline (with but one exception, in which only at the beginning of experiment the temperature rose one-eighth of a degree), was always observed; in one instance the temperature fell as low as 34.5° C. Iced saline caused a fall in pressure when the first flow reached the heart; however, the pressure soon rose; in using the warm solution, this preliminary fall was in no instance noted. Local cooling, or even freezing of a part likely to contribute to shock, would, to the extent it blocked afferent impulses, prevent the contributing effect to shock; but reaction is always to be considered, and, besides, our observations are more especially directed to the systemic effect rather than to the local.

PHYSICAL CONDITION

Animals in impaired health, insufficiently nourished, too young or too old, were correspondingly bad subjects; even the laboratory servant's estimate of any given animal, as to its value for a protracted experiment, was usually well taken. Pregnancy did not seem to impair the resistance, but this statement rests largely upon personal judgment. In bad subjects and after shock had been fairly induced, the reflexes were never so acute, vasomotor actions were sluggish and ineffectual, and when the blood-pressure declined, it was, as a rule, slowly or never regained. Animals having firm tissues seem to have greater resistance. Gross tissue resistance proved to be a very fair index to vital resistance.

ANÆSTHESIA

So important a *rôle* was played by the anæsthesia that it was necessary to be constantly watchful, and when sufficient assistance could be obtained we supplied a special anæsthetizer. The respirations in over-anæsthesia became gradually more shallow and slower, and if the anæsthetic was continued, would fail suddenly. The blood-pressure *pari passu* gradually fell. Upon removing the ether, both would rise, much as they fell. When the animal was about to come out of the anæsthesia, respirations gradually deepened and quickened and the blood-pressure curve became irregular,— precisely the reverse process to the entrance into the state of anæsthesia too profound. The respiratory indications were usually in advance of any other symptom in foretelling the tendency of the anæsthesia. The effect upon respiration was so constantly in advance of other effects—for example, that upon the circulation—that the latter was habitually neglected ; and in no instance, excepting in several cases of chloroform anæsthesia, was there a sudden failure of the heart. If the animals were allowed to partially recover from the anæsthesia, care was necessary in reducing them again to surgical anæsthesia, as the tendency to hyperpnœa attends this state, and an excess of the anæsthetic was likely to be inhaled. The exhausting over-respiratory efforts predispose the respiratory centre to failure, and special care was always necessary at this point. The more the animal had been subjected to operative procedure, the more readily the respirations failed under the above conditions. In abdominal operations, contact with the peritoneum caused an increased, though irregular, respiratory rhythm, and this was occasionally mistaken for indications for more of the anæsthetic. Chloroform required much greater care than ether ; ether in no instance caused sudden cardiac arrest ; chloroform, three times, each time early in the inhalation and before surgical anæsthesia had been induced. When accordingly more of the anæsthetic was administered, a profound depression was produced. Chloroform proved to be more toxic than ether.

HEMORRHAGE

Loss of blood always predisposed to shock, and when considerable, even if it caused but little depression in the blood-pressure, the animal did not withstand so severe or so protracted an operation. Hemorrhage from the large venous trunks caused the most profound effect upon the blood-pressure, because the quantity of blood supplied to the heart was immediately diminished, while if the hemorrhage was arterial, the income of blood was not so suddenly diminished. The output of the heart does not depend at all upon the height of arterial pressure, but is in direct proportion to the venous pressure. Inasmuch as the venous pressure is more directly lowered by venous hemorrhage, the observed facts are in accord with the observations of Donaldson on the output of the heart. Dogs having enlarged thyroids have usually disproportionately large carotid arteries. Their blood seems to clot less readily, hemorrhage is everywhere more free, and there is less tendency to its spontaneous arrest. Hemorrhage is especially free in the neck. This observation, so far as I am aware, has not found its way into surgical literature, at least; and if true, as in the case of dogs there is abundant evidence, it may serve as an explanation for the usually troublesome hemorrhage sometimes encountered in thyroidectomy. It has been said by able clinicians that shock is only another name for hemorrhage; but in this series of observations it is apparent that hemorrhage may be merely one factor in the production of shock. In the present series, unless the fact is particularly mentioned, there has been no consequential hemorrhage. Respirations are always accelerated and deepened in profuse hemorrhage. A given amount of loss of blood when the animal has been already reduced by surgical shock caused more depression than an equal amount earlier, when the compensating vasomotor mechanism was more active.

VASOMOTOR CHANGES

There is sufficient evidence at hand to establish at least a high degree of probability that the shock in operations on the splanchnic area is largely caused by disturbance of the local splanchnic vasomotor mechanism. The experiments of Mall show that the splanchnic nerves are vein-nerves and control this large and spacious vascular area. Every experiment in this area gave evidence of the dilatation of the vessels controlled by these nerves, and the decline of the pressure occurred *pari passu* with this dilatation. A water-manometer inserted towards the heart in the splanchnic vein showed a rise in pressure, while the central was falling; there being no valves in the veins of this area, the manometer probably correctly registered the portal pressure. The entire splanchnic area, excepting the distal part of the splenic vein, was left unimpaired. The spleen was either extirpated or its arterial supply clamped. The veins were much more dilated than the arteries; in fact, towards the close of the experiment, when the veins were much engorged, the arteries were comparatively small. While the splanchnic vasomotor mechanism played an important part, that it is not the whole cause rests on the following evidence: Clamping the thoracic aorta, or all the splanchnic arteries retroperitoneally, to effectually remove this area of vessels as a factor, on extensive and continuous manipulation and operation, the intestines became paler than normal and the veins comparatively empty. However, death was, under these conditions, caused from shock by mechanical injury of this area, though not so readily as in cases in which the splanchnic arteries had not been thus clamped. The blood-pressure did not suffer decline as under normal conditions, but the respirations were affected no less than under normal conditions, and after continued manipulation of the intestines the respirations became gradually more shallow and usually failed before the heart. The latter grew gradually weaker, and finally failed, the resulting death being independent of the vasomotor *splanchnic* factor. It must be observed, however, that the animal with excluded splanchnic circulation endures much more splanchnic injury. The great shock caused by injuries of the male genital organs is probably due largely to vasomotor dilatation of the splanchnic area. However, even in cases in which the splanchnic vasomotor mechanism had been excluded there probably were afferent impulses produced by injuries upon the splanchnic area which reached the higher vasomotor centres, and thus influenced the calibers of the blood-vessels in areas other than the splanchnic. It is then safe to say that the "excluded splanchnic circulation" experiments only prove that the splanchnic vasomotor factor plays but a part even in such injuries as involve its own area alone; there is no evidence tending to show that in operations in areas of the body other than the splanchnic and genito-urinary, the splanchnic factor plays a special part, probably not much more of a part than in any other area of like vascular capacity. Autopsies, in experiments in which this area was not involved, did not reveal a condition of vascular distention in this area different from that of other areas.

In operations upon the somatic area,—*e.g.*, the extremities,—the effect upon the vasomotor mechanism was most certainly shown in the production of Traube-Hering curves,—undulating movements in the central blood-pressure, due probably to rhythmic vasomotor action. Early in an experiment, while the animal was fresh, these and other vasomotor effects—*e.g*, manometers in peripheral area, or the water-manometers in the abdominal vessels showing alterations in pressure at variance as to direction of the curve with the central blood-pressure, or showing a disproportionate alteration in the same direction, or finally showing a variance in point of the time of the alteration in its curve—were readily produced; but after the animal had become somewhat exhausted, or in a weakly animal even earlier, a vasomotor effect could scarcely be produced.

Then, while in the first injury, or at least in one of the early injuries or stimulations of a nerve-trunk, a rise in blood-pressure was always produced, after a number of repetitions a condition was reached in which no effect was produced, and later, on stimulation, an actual fall, without a preliminary rise, occurred.

In the early experiments, while the nerve-fibre was fresh, after the uniform early rise, there would often later in the adjustment of the curve be a lower pressure than before the experiment was made. However, in cases in which the stellate ganglia had been removed prior to the stimulation of a peripheral *nerve-trunk*, such stimulation then applied produced neither a rise nor a fall. These several results, so many times obtained, are taken as evidence of a vasoconstrictor mechanism or action and a vasodilator mechanism or action, or, in other words, a pressor and a depressor action, and that the former is exhausted more readily than the latter.

It is well known that if a nerve-trunk be subjected to extreme cooling, say to 0° C., and is then stimulated, in most instances the stimulation produces a fall, while before such cooling a rise will be produced, the conditions for experiment being favorable.

In regeneration of nerves and under other circumstances it has been shown that the vasoconstrictor or pressor action is more readily exhausted than is the vasodilator or depressor action. Whether these are separate mechanisms, or whether there is but one mechanism responding differently under conditions we can but feebly understand, does not in the least concern the present inquiry.

After the animal had been reduced to the condition referred to,—viz., stimulation or injury producing a primary fall in pressure, instead of the primary rise, as at the beginning of the experiment when the animal was fresh,—it was found to be practically impossible to raise it by therapeutic measures. The venous pressure had sunk so low that the heart received but little blood upon which to act; the vasoconstrictor mechanism was wholly inactive, except in response to most heroic stimulation, and when this was applied the temporary gain was followed directly by a still greater loss in blood-pressure.

A number of extensive dissections were made to observe the condition of the vessels at this stage. The arteries were quite empty, the tissues pale, but the venous trunks everywhere were full, alike in the somatic and in the splanchnic areas. There was a manifest transference of the greater bulk of the blood from the arteries and capillaries to the veins. Making such dissections for this and other purposes, the relative proportion of blood in the arteries and capillaries and that in the venous trunks gave a rough estimate of the depth of the shock at that time.

The blood was to a corresponding degree dark colored. There was a strong suggestion of a loss of venous tone, out of proportion to at least the superficial evidence of lost arterial tone.

If, after such a degree of shock had been produced, with such general venous engorgement, such arterial and capillary depletion and such loss of vasoconstrictor or pressor action, an infusion of warm saline solution was allowed to flow into a vein, the blood-pressures, as will be later described more fully, all rose; all the peripheral, venous and arterial, portal and cephalic, rose within a few seconds of the beginning of the saline flow.

The heart-beats were strong and full, and their uncountable rapidity was reduced to a normal, and in many cases to a slower than normal rate; the output of the heart was enormously increased, as indicated by the sweeping strokes in the tracings. Now, if the animal had been reduced too nearly to death, these long and slow beats, with their enormous output of fluid with each stroke, became slower, and the blood-pressure did not rise. The beats continued for some time, then suddenly ceased. On the other hand, if the animal had not been reduced too nearly to death, the renewed cardiac action not only raised the blood-pressure, but, after a time, vasomotor actions frequently reappeared. The reappearance of the lost vasomotor tone may be due to the increased amount of blood sent to its centre with each of the strong heart-beats, or to recovery from rest, or to both. What has the saline done? It has increased the venous pressure, which in turn filled the heart; this in its turn beat strongly and sent out larger quantities of saline blood, which in turn fed the exhausted and starving centres, and carried the overcharged blood to the lungs for the respiratory changes. Like benefits, though to much less degree and less positive, were obtained by inclining the head of the dog-board downward, thereby invoking the aid of gravity.

These several groups of observations, so many times repeated, would tend to show how great a factor is the exhaustion of the vasomotor mechanism, notably the vasoconstrictor or pressor actions. The saline experiments show well how much more competent is the heart than the vasomotor mechanism to withstand the causes of shock. There are a number of exceptions, and many facts bearing on the important cardiac factor, which will be discussed under another heading.

In a number of vasomotor phenomena observed it was apparent that there are regional vasomotor actions quite independent of each other,— e.g., while the area of a leg may indicate a decline in blood-pressure, the portal may simultaneously show a rise, and vice versa. These observations of regional actions did not appear later in the experiments, when the animal had become well exhausted.

A very considerable number of tracings of the cephalic circulation did not seem to indicate any appreciable vasomotor action in this region, so that, so far as the brain was concerned, its circulation was influenced by vasomotor disturbances only indirectly, as such disturbances affected the general blood-pressure, and so that of the brain.

In operations upon the cerebral cortex, in operations involving the brain-tissue, and especially in such procedures as caused injury, either directly or indirectly, upon the base, involving the medulla in particular, striking vasomotor effects were produced.

Operations whereby the fourth ventricle was exposed bloodlessly—at least nearly so—were attended by most profound vasomotor disturbance, and the vasoconstrictor or pressure action was very early lost by mere exposure of this region to the air. The general blood-pressure gradually declined in the usual way after vasomotor break-down.

There is no portion of the circulatory apparatus so delicate, whose equilibrium is so easily disturbed, and whose connection with all the parts of the body is so minute, as the vasomotor. A priori, it would be the most readily disturbed, as well as most readily exhausted, and to such conclusions do our observations lead us.

The more richly supplied with vasomotor nerves was a given area, the more rapidly was the vasomotor mechanism exhausted when such area was subjected to injury. This principle was abundantly illustrated in the experiments upon the splanchnic area.

It may also be stated that a given injury produces shock in proportion to its severity and the duration of its application. Whether such injury produces a rise or a fall of pressure, or even Traube curves, it contributes its certain quota of the total amount of damage necessary to the final production of vasomotor break-down.

CARDIAC CHANGES

The heart is the base of support of the blood-pressure, and any interference with its action at once causes marked changes in the pressure. The output of the heart is in direct ratio to the pressure of the vena cava, and not at all to the height of the aortic blood-pressure. The venous pressure, then, determines the heart's output, and the venous pressure is, in a good measure, dependent upon the force and frequency of the heart-beats, together with the necessary vascular tone, which is under control of the vasomotor nerves. If, then, the area of peripheral resistance be diminished, the venous pressure will be diminished, and so the output of the heart will be diminished. Supposing the peripheral resistance remained unchanged, if the heart's action be diminished in force or frequency, or both, the venous pressure will fall pari passu with the central; and, again, with the lowered venous pressure, the output of the heart would be diminished. The cardiac action and the peripheral resistance remaining the same, venous hemorrhage or mechanical interference with venous pressure, the venous pressure will be lowered and the output of the heart diminished. Other things remaining the same, an increased output of the heart causes a rise in the pressure, and *vice versa*. This was repeatedly shown by the effect of the infusion of normal salt solution. The efficiency of the heart-action may be diminished by an over-stimulation, or by a continued use of stimulus. Exhaustion follows such overaction. There may be over-stimulation of the augmentary mechanism. The heart-action may also be interfered with by the excitation of the inhibitory impulses, as in operations involving the inhibitory area in the larynx, brain, etc., as heretofore described, often causing a staggering fall in pressure. The heart-action is also greatly dependent upon efficient respiratory action, as is shown in a number of instances in which respirations had failed, and the heart, beating on in sweeping strokes, soon stopped. Also, upon producing artificial respiration, the heart, after a quiescence of as much as fifty seconds in one case, began action again. The diminished temperature of the blood, previously referred to, might be a cause of impaired heart-action. Upon this point we are not prepared to offer any new evidence. In the experiments in which the cardiac branches of the stellate ganglia had been severed previous to making tests, the heart beat on slowly, and although death was delayed, it occurred while the heart was beating slowly and regularly. This is evidence that the rapidity of the beat late in an experiment, when shock is very profound, was due to impulses from the higher cardiac centres, in an effort to regain the lost blood-pressure, and not due to the diminished amount of blood supplied to the heart. The circulatory regulating apparatus in restoring a lost pressure sends out efferent impulses, not only to the heart through the cardio-accelerating apparatus, but also to the blood-vessels (peripheral resistance) through the vasoconstrictor apparatus. Both usually act simultaneously. Now, in the condition of profound shock, the accelerated cardiac action is taken as evidence of the effort of the regulating centres to recover the lost blood-pressure; but the vasoconstrictor or pressor mechanism is not active; it is, therefore, probably paralyzed, or at least seriously impaired.

The splendid action of the heart under these conditions, on the infusion of normal salt solution, proves it to be well capable of further action. The general property of heart-muscle fibre, whereby even the mammalian heart may be kept beating for hours after complete removal from the body, gives assurance of its endurance. The amount of evidence at hand leads to the conclusion that the heart's action is the last of the great vital functions to fail.

Respiration.—In one hundred and three of the experiments in which the exact manner of death was recorded, or in which, in the course of experiments, either the heart or the respiration failed first, respiration alone failed in ninety, the heart alone in four, and both simultaneously in nine. In many instances the heart was beating strongly and the blood-pressure was fair at the time respiration failed. Artificial respiration was frequently required during the course of the experiments. The greater the extent of the dissection, and especially if dissection had been made in the thorax or the abdomen, the more readily respiration became exhausted. In bloodless amputations of the hip-joints and other mutilating experiments, respiratory failure occurred first. Almost every injury causing any effect on the circulation causes respiratory changes, usually more striking than the vascular, and in many experiments, notably in the splanchnic area, respirations were more sensitive to irritation than was the circulation. In traumatisms of the brain the respirations were strikingly more affected than the circulation, and the immediate cause of sudden death from traumatism of the brain was in almost every instance failure of respiration. In one gunshot wound, not even touching the medulla, but imparting to it the percussion of the ball, death was caused by respiratory failure. In gunshot wounds of the chest, when large vessels of the heart were not penetrated, death was caused by respiratory failure. In laryngeal operations and injuries respirations were very easily inhibited. In almost every instance of dangerous anæsthesia, the respirations were most affected, and frequently stopped suddenly. The heart could be depended upon to continue beating long enough to establish artificial respiration. With the amount of evidence this research has furnished bearing on the sensitiveness of the respiratory mechanism, we are led to estimate it as one of the important factors in the production of shock. There remains a residuum of phenomena whose explanation has not been attempted.

POST-MORTEM APPEARANCES

Autopsies were made in a large number of the animals, and, generally speaking, the following conditions prevailed : the large venous trunks were full, sometimes enormously engorged, the arteries empty, the veins of the splanchnic area not more distended than those of the somatic, unless the experiment had included some procedure in the splanchnic ; the left ventricle and the left auricle were empty or nearly so ; the right auricle usually contained some blood, the right ventricle little or none ; the lungs were anæmic ; pulmonary vessels empty ; tissues of the brain and of the somatic area anæmic ; liver usually engorged ; spleen and kidneys somewhat less so. The disposition of the mass of blood was in a number of cases observed by making the necessary dissection during and before death.

ON THE PREVENTION OF SHOCK
HEMORRHAGE

In attempting to operate without producing shock it was necessary to prevent as far as possible the loss of blood; even though the loss of some blood did not at the time produce an appreciable effect, if these small losses were repeated, the animal the more readily succumbed to shock. On diminishing the quantity of fluid in the vessels an additional stress is placed upon the vasomotor centre in maintaining a given mean blood-pressure. The reason for this is quite apparent. Every such stress placed upon this centre diminishes the total amount of functional reserve it may have, and hence, as experience has abundantly proved, vasomotor tone in the subsequent course of experiment in cases of hemorrhage is the more readily lost the more the hemorrhage has been. It is most important to prevent hemorrhage from the large venous trunks. On account of the alterations in the intrathoracic pressure due to respiration, the recurring negative pressure in the great venous trunks of the neck renders air emboli possible, though not probable. Not probable on account of the valvular action of the collapsible venous wall. Although the large veins were frequently partially or entirely severed, air in but two instances entered. Even then, the importance of air embolism is in proportion to the amount of air and the height of the blood-pressure. In a low blood-pressure, when there is also low venous pressure as well, the output of the heart is small; its chambers are only partially filled, and the air emboli are not so readily dislodged from the chambers. It must be remembered also that, since twenty-one volumes per cent. of the air is oxygen, the venous blood will soon take up the oxygen, or twenty-one per cent. of the embolus. The law of the partial pressure of gases will operate to diffuse the remainder, provided that death does not quickly ensue. A number of animals were killed by introducing air into the veins, and a very considerable amount was always required. In experiments on puppies, and very old, as well as ill-fed animals, the greatest care was necessary to avoid hemorrhage.

THERAPEUTIC MEASURES

Atropine, hypodermically administered, was an efficient protection against cardiac inhibition in operations in the "inhibition area," in the larynx, and in such operations as might cause mechanical stimulation of the vagi. Ten observations on preliminary injection of atropine to prevent splanchnic shock leave us in doubt as to its efficiency; but, on the whole, the evidence would seem to show that shock was diminished. The blood-pressure curve was very irregular, on account of the loss of the

10

" vagal" influence. The blood showed an unusual tendency to clot in the atropine experiments.

Cocaine, hypodermically, guards the heart against cardiac inhibition almost as efficiently as does atropine. The reflex impulses in the laryngeal area of inhibition are wholly prevented by a one-half per cent. solution locally applied; its action is pronounced in thirty seconds. Cocaine applied to nerve-tissue entirely inhibits its function. Nerve-trunks may be physiologically "blocked" by cocaine, so that neither afferent nor efferent impulses of any kind can pass : hence the area so protected cannot, on any operation or injury, cause the slightest degree of shock. Applied locally upon the peritoneum, it prevents the respiratory effects of manipulation. In a series of experiments on cocaine administered hypodermically to prevent splanchnic shock, there is strong evidence as to its efficiency. The blood-pressure curves under cocaine were irregular.

As to morphine and alcohol, our observations were rather negative.

METHODS OF OPERATING

Tearing, manipulating, stretching, forced dissection, all tend to produce shock, more than the use of sharp instruments and gentle manipulation. The contrast is most striking in direct observations on nerve-trunks. Naturally, the more important the function of the part under operation, the greater the necessity for such care.

ANÆSTHESIA

Reference to ideal anæsthesia has been made. Animals to which, while inducing anæsthesia, an overdose had been given, did not endure a prolonged experiment, and not only was it more difficult to maintain an even anæsthesia afterwards, but the animal also showed a marked tendency to recurring respiratory failures. Over-anæsthesia rendered the animals subject to early collapse and decidedly less capable of enduring a protracted experiment. Respiration usually failed first, though the blood-pressure gradually declined with the shortening of the respiratory stroke. On removal of the anæsthesia, the blood-pressure and the respiration were generally restored in consonance. Chloroform caused sudden arrest of the respiration more frequently than ether. In three cases the heart was suddenly arrested. There can be no doubt that chloroform is much more toxic than ether. Much greater care was necessary with chloroform than with ether.

During the performance of operations on such fields as produce an increased respiratory action, considerable care was necessary to prevent excessive inhalation of the anæsthetic by the animal, notably in abdominal, axillary, and anal operations.

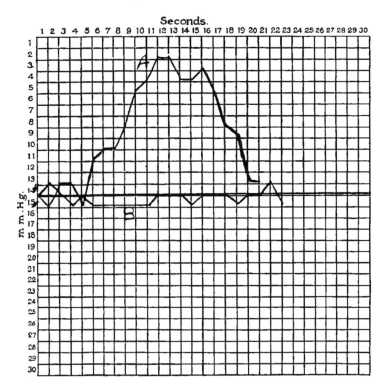

A, burning foot as control experiment. *B*, burning of opposite foot after injecting cocaine in the sciatic and anterior crural nerves. Ten experiments.

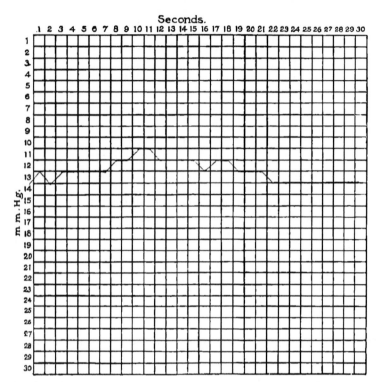

Exposing and manipulating the intestines after administering cocaine. Twenty experiments.

TEMPERATURE

So far as could be judged, less shock was produced when warm solutions were used than when cold; when the laboratory was warm than when near the freezing-point. Moist heat, protecting the tissues, certainly lessened the amount of local irritation, and hence the shock. This was notably observed in experiments on the intestines and the brain.

POSTURE

Especially later in the experiments, posture was of considerable importance; most notably when there had been consequential hemorrhage. The blood-pressure always rose when the dog-board was tilted head downward, and fell when the board was tilted in the opposite direction. When from the effect of the experiment or the anæsthetic, or both, the vasomotor mechanism lost its activity, compensation would not readily take place, especially when there was massing of blood in the splanchnic area. After abdominal operations the effect of inclination was more marked.

REGIONS OF THE BODY

In operations upon the extremities, if performed bloodlessly and if the nerve-trunks had been subjected to a cocaine " block," no shock could be produced. No afferent impulses could pass the " block," and amputation of a leg caused no more effect than did cutting the hair.

In operations upon the chest, especially in the axillary region, care was necessary not to cause undue mechanical irritation of the nerve-fibres; for it was observed that in rough axillary and chest dissections there was a marked tendency to respiratory failure. The respiratory centre, being in direct connection with the nerves of this region, was much affected and finally broke down, as a result of the afferent impulses caused by the dissections. These observations have a bearing upon the dissection of the axillary space in carcinoma, etc. The same tendency to respiratory failure was observed in operations within the sphere of the nerve-supply of the general respiratory mechanism, which, of course, included the entire chest.

In abdominal procedures, if the omentum was made to cover the viscera, thereby preventing direct contact with them, there was very much less shock. The omentum is the antithesis of the intestines. Its manipulation does not produce vasomotor and respiratory disturbance. The less the manipulation and exposure the less the shock. Dragging on the mesentery caused mechanical stimulation of the splanchnic trunks, and a decline in blood-pressure soon followed. Not only this, but traction on the mesentery caused a mechanical interference with the return supply of blood in the large venous trunk, and so produced marked alteration in the blood-pressure. In operations upon the gall-bladder and liver it was

necessary to be cautious as to traction, on account of the great fluctuation
in blood-pressure likely to be caused by mechanical interference with the
large venous trunks.

In operating upon the kidneys, if the peritoneum was avoided, but
little effect was noted. The most dangerous area was found to be in the
region of the duodenum, pylorus, and gall-bladder. Here the great
venous trunks, the diaphragm, the splanchnic nerves, are accessible; all
may be important factors in producing shock. Dragging unduly upon the
pylorus caused very marked alterations in blood-pressure; manipulation
upon the diaphragm caused the most striking respiratory alterations, and,
if continued, was likely to induce a break-down of the respiratory centre.
The least dangerous area is the pelvic peritoneum and its viscera.

The uterus, tubes, and ovaries contribute but little to shock, even when
they are severely crushed and torn. The severity of shock induced in
abdominal operations is in indirect ratio to the distance from the pelvis.
Injuries of the large intestines produce much less depression of the blood-
pressure than those of the small intestines; injuries of the stomach, about
the same as the small intestines. In intrathoracic operations the greatest
care was necessary in touching, even gently, the heart or the great venous
trunks; the arterial trunks are of much less importance. The slightest
contact with the heart or the large venous trunks causes the most profound alterations in the heart-action and the blood-pressure. The same
may occur in indirect pressure upon the heart through the diaphragm, in
manipulations from the abdominal side of the diaphragm.

In operations on the male genital organs great gentleness is necessary,
to avoid the profound fall in blood-pressure so likely to occur. Preliminary injection of cocaine at the seat of operation almost wholly prevented such effects.

ON THE TREATMENT OF SHOCK

Aqueous extract of the supra-renal capsules of sheep was tested. There
was, on its administration, an immediate and marked rise in blood-pressure, but the pressure was not at all sustained and the fall was as rapid as
the rise.

Strychnine sulphate caused a marked, but later rise in all the pressures,
and the rise was well sustained.

Artificial respiration is of undoubted importance, and we have in our
collection many records of the remarkable effect of artificial respiration
on the heart-action and the blood-pressure. The blood in shock is insufficiently supplied with oxygen. An increased supply of oxygen is
urgently demanded.

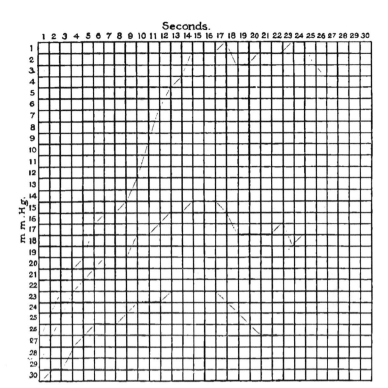

saline transfusion. Ten experiments. Upper curve represents the splanchnic venous pressure, the middle the central blood-pressure, and the lower the peripheral venous. Each square represents *two millimetres of mercury and five seconds of time*. *P*, Portal pressure; *A*, arterial pressure; *V*, venous pressure.

Intravenous infusion of normal saline solution causes, in the first instance, an increase in the venous pressure in the vena cava, and consequently the output of the heart is at once increased, the amplitude of the strokes lengthened; the chambers being full of blood, contractions in consequence become more forcible and the blood-pressure rises several beats after the beginning of the saline flow. The peripheral venous, the vena cava, portal, peripheral cephalic, and central pressures all rose together. The rise of the vena cava and peripheral venous pressures appeared first, and last and after the lapse of some time, the portal rose, but rose proportionately higher, if later, than the other pressures. If a small quantity of saline solution was injected, the rise was likely not sustained. In a ten-kilo dog as much as seven hundred and fifty cubic centimetres were injected continuously and the elevated pressure was finally sustained. Injections of quantities of fifty cubic centimetres at intervals, as needed, served the purpose well. Quantities up to twice the amount of blood calculated to be in the animal have been given before the pressure was sustained. After this considerable injection the wounds everywhere began oozing and the tissues seemed wet. Hemorrhages that had been insignificant recurred vigorously. Various temperatures, ranging from 18° to 53° C., were employed, and within a reasonable range the effect seemed to be the same. The value of these injections is apparently wholly mechanical. In a small dog five hundred cubic centimetres at 53° C. were injected while the rectal temperature was being taken, and it was found to have increased the temperature but one-fifth of a degree. Hemorrhage, after injection of considerable saline solution, shows little tendency to spontaneous arrest or clotting. The combination of small and frequently repeated hypodermic injections of strychnine, together with saline infusion, is most effectual. Strychnine alone causes acceleration of the heart with shorter beats. Saline solution alone causes a more forcible, but a long, sweeping beat. The combination of both produces a more sustained effect. Small doses more frequently repeated, producing an effect similar to the steady increment of a continuously flowing stream of saline, have appeared to produce the best results.

Over-stimulation was followed later by a greater depression.

The application of heat was of benefit. Its application in the abdominal cavity seemed, on the whole, of doubtful value. While it exercised at first a beneficial influence upon respiration and the blood-pressure, there seemed to be a later decline so great as to make the procedure of doubtful value.

The same may be said of hot water in the stomach and rectum, though in these observations the later decline was less marked.

Filling the stomach with hot water, or passing a stream through the stomach, caused a marked improvement in respiration ; but the gain in blood-pressure was not marked. In several instances, however, the blood-pressure was raised.

POSTURE.—Inclining the animal with head downward was of very considerable benefit. The blood-pressure immediately rose and the respirations were increased. The more nearly the vasomotor mechanism had reached its final break-down, the more marked were the effects of posture upon the circulation. In some experiments one hind leg during the development of shock was kept elevated, the other in the horizontal posture. While the large veins in the latter, towards the close of the experiment, were distended with blood, the former were quite empty.

CONCLUSIONS

The following conclusions seem to be justified by a consideration of the data afforded by this research, and what follows relates only to what has preceded. As a result of injury or operation, there may be immediate or later depression or death.

Immediate depression or *death* from injuries or operations was due to one or more of the following factors :

1. CARDIAC.
2. RESPIRATORY.
3. HEMORRHAGIC.
4. VASOMOTOR.

The phenomena produced by the operation of these factors have so many points in common that they naturally fall into a group and would seem to be appropriately designated as *collapse.*

Later depression or death from injuries or operations and due to a *vasomotor* impairment or break-down as the essential factor may be appropriately designated *shock.*

FACTORS PRODUCING COLLAPSE

CARDIAC

The cardiac factor may be due to—

(*a*) Injury of the heart-muscle, as in gunshot and stab wounds, etc.

(*b*) Injury or operation involving the vagi or certain of their branches, —*e.g.*, a direct blow upon the vagus, intra-laryngeal manipulation, etc.

(*c*) Injuries of the cardiac centres in the medulla,—*e g.*, great intracranial pressure, gunshot wounds involving the medulla, crushing of the skull, etc.

(*d*) Paralysis of the heart-muscle by excessive anæsthesia, notably by pushing chloroform or by administering an injection of the same into a vein ; the heart-muscle fibre is thus readily paralyzed.

RESPIRATORY

This factor may be introduced by—

(a) Injuries or operations upon the thorax, destroying its integrity, thereby producing collapse of the lungs.

(b) Injuries or operations involving the respiratory nerve-mechanism, —e.g., destruction of phrenic nerves in continuity, cross-lesion of the spinal cord above the fourth cervical vertebra, mechanic stimulation of the superior laryngeal.

(c) Injuries or operations affecting the respiratory centres in the medulla as well as the higher respiratory centres at the base of the cerebrum,— e.g., excessive intracranial pressure, gunshot wounds, compound fractures, etc.

(d) Mechanic interference with the exchange of gases.

(e) Excessive anæsthesia, whereby respiratory action is arrested.

HEMORRHAGE

Excessive hemorrhage especially from the large venous trunks may cause collapse.

VASOMOTOR

The vasomotor centres are continually in action, and exercise thereby a continuous pressure upon the contents of the blood-vessels. This continuous pressure is known as vascular tone. A cross-section of the spinal cord at any given point produces loss of this vascular tone, thereby permitting the vessels to passively dilate to their utmost, and if the lesion involves a sufficiently large area of vessels, collapse may be produced by the accumulation of blood within this area of vascular dilatation, being practically equivalent to a hemorrhage. The same may be said of sections of such nerves as the splanchnics,—nerves that control a capacious vascular area. It is only this phase of vasomotor influence that may be particularly operative in producing collapse. The vasomotor nerve-mechanism may also be paralyzed by an excess of anæsthetic, especially by chloroform.

SURGICAL SHOCK

Surgical shock is mainly due to impairment or break-down of the vasomotor mechanism. All the factors referred to as causing collapse may add to shock. The evidence from which this conclusion has been reached is the following :

Nerve-fibres when subjected to stimulation, mechanic or otherwise, may cause either an increase or a decrease in the blood-pressure. The presence of the so-called "pressor" nerve-mechanism is well established, as is also the presence of the so-called "depressor," and these "pressor" and "depressor" nerve-mechanisms have a representation in the nerve-trunks. Now, mechanical stimulation of sensory nerves when both the animal and the nerve are fresh produces a rise in blood-pressure, due largely to a stimulation of the "pressor" mechanism. On repeating the stimulation of the same nerve each subsequent repetition is attended by a diminished pressor action, and finally no effect is produced upon the blood-pressure. On the contrary, there may be a fall in blood-pressure after still further stimulation,—that is to say, the depressor action alone may appear. Not only is this true with reference to repeated stimulation of the same nerve, but after repeated stimulation of other nerves there will be finally a condition produced in which no stimulation even of a fresh nerve-trunk will produce a rise in the blood-pressure; in other words, it would seem that the pressor action may be exhausted in a given nerve-trunk by repeated stimulation of different nerve-trunks, the latter being general, the former local, in the sphere of its influence. It has been almost universally noted that the development of shock increases *pari passu* with the failure of the pressor action,—that is to say, so long as the blood-pressure remained high or fairly high it would be found that the pressor action could be brought out on stimulation of a fresh nerve-trunk. On the contrary, after a sufficient number of stimulations there would be a marked lowering in the blood-pressure, and at the same time the pressor action became greatly impaired or entirely abolished. As nearly as could be estimated, the depth of the shock was in proportion to the failure of the pressor action. And when a given nerve, or when the entire animal had been reduced to such a condition, further stimulation in many cases produced a primary fall in blood-pressure, and generally without a later compensatory rise. It has been pointed out by others that when a nerve-trunk is severed and repair has taken place a stimulation applied early to the repaired nerve-trunk will produce a primary fall in blood-pressure, but such stimulation applied later will produce a rise in the blood-pressure,—that is to say, the depressors repair earlier than do the pressors. If a nerve-trunk be reduced in temperature to a point at or below zero, then the distal end stimulated, there will be a lowering of the blood-pressure,—a depressor action. In this way the two actions, the pressor and the depressor, are separated. It would appear, then, that the pressor action is brought out on first stimulation or early stimulation, and that the pressor action is more readily exhausted than is the depressor action. The physiologic importance of the pressor action is too apparent to be necessary to be entered into here, and the helplessness of the circulatory apparatus, when its pressor action has been lost, must be complete. In this condition of lost or impaired pressor action the blood is transferred to the venous side, notably to the large venous trunks, all of which are dilated. In a given tracing the volume of blood put out of the heart with each beat is indicated by the stroke of the writing style of the manometer. This stroke in every case was lessened as shock was developed. When the dog-board was inclined with the head of the animal downward, the blood-pressure underwent a relatively great rise, suggesting the emptying of the blood from the engorged toneless vessels in which it had accumulated. The usual physiologic compensation in changing of posture was not manifested in profound shock. In a fresh animal, and when the circulatory apparatus is normal, if an animal be suspended feet downward, there will be an immediate fall in blood-pressure, but directly after there will follow a compensatory rise, so that in a short time the blood-pressure will have wholly regained its normal level and sometimes temporarily rise even above the normal. This has been demonstrated to be due largely to the vasomotor mechanism regulating the caliber of the vessels. In the state of shock (and the phenomena are almost in direct ratio to the depth of the shock), when an animal is inclined hind feet downward, thereby causing a fall in blood-pressure, there is but little, if any, compensatory rise, and in some cases of extreme shock there was no compensatory rise at all. The amount of compensation was believed to be in proportion to the depth of the shock,—that is to say, the blood-vessels were so nearly toneless, and were so far out of the control of the vasomotor mechanism that their caliber was not altered in the normal process of compensation, supporting the observations just made that in profound shock there is a loss, partial or complete, of pressor action. It was further observed that if an animal were allowed to partially recover from its condition of profound shock, and then inverted, there was an inauguration of a compensation seemingly in proportion to the extent of its recovery from shock. It would then seem that in the fatal cases of shock and in the conditions of most profound shock there is a condition of either total paralysis of the vessel-walls or its equivalent.

Negatively, that the heart is not exhausted, and that it is not the principal factor in the production of shock, rest on the following evidence:

The heart-muscle itself is functionally characterized by its capacity for work, as athletic exercises, physical endurance tests, etc., prove. An experimental mammalian heart has been made to beat as long as five hours after its removal from the body, having as its circulating medium defibrinated blood. Now, in the condition of shock there is distinctly less work for the heart to do, and the more profound the shock the less the work, estimated by the amount of blood the heart puts out. But it may be asserted that the heart beats with increased frequency *pari passu* with the development of shock, and its "machinery running empty" runs away with itself, and so becomes exhausted. But in the experiments in which the cardio-accelerator fibres had been previously severed, and the heart in consequence beat slowly until death, a descending curve in the blood-pressure showing the final vasomotor break-down was practically the same as in the experiments in which the heart ran rapidly until the end. Finally, when the heart had been allowed to so "run down" almost to the point of ceasing, and normal salt-solution was allowed to flow into a vein, causing thereby increased venous pressure, hence supplying the heart with more fluid, the heart was at once reduced to slow and full beats, showing that it was not exhausted, and that it beat rapidly because it did not have enough to do rather than that it was overworked. In this condition each heart-beat produced a long stroke of the manometer, the extreme length of the stroke being caused by the combined effect of throwing maximum quantities of blood into the aorta with each beat and the toneless condition of the vessels, so that on account of this toneless condition the increased pressure produced by each individual beat was not sustained, and therefore the excursions were extremely long. After a number of these sweeping strokes had been executed and the vasomotor tone was again regained, the general blood-pressure rose, the strokes became more rapid and shorter, approaching the normal, and at this time vasomotor curves again reappeared. Before this, while by the extraordinary length of the strokes the toneless condition of the vessels was announced, no vasomotor curves appeared. In cases in which the vasomotor tone was not regained, the long sweeping strokes went stalking along until the heart failed to contract and death occurred in diastole.

The cardiac factor no doubt plays a part in the production of shocks because a normal cardiac action is essential to the maintenance of the normal blood-pressure, and it is highly probable that the causes producing the depression or break-down in the vasomotor system also cause an impairment of cardiac action, as has been pointed out in referring to the cardiac action in shock under that head in the summary. The respiratory factor is also an important one, inasmuch as the respiratory action was noted to have been altered in almost every injury or operation that may produce shock, and the respiratory action is of great importance, equal to or greater than that of the cardiac. Impaired respiratory action would lead to impaired gaseous exchange in the lungs, thereby the vasomotor as well as all other functions would be impaired. But shock was produced in all the experiments in which a sufficient exchange of gases was provided by means of artificial respiration.

The influence of hemorrhage on the production of the shock was very marked. It need only be stated to be accepted that the vasomotor system, in adapting the caliber of the vascular system to the needs of the circulatory apparatus so as to maintain a mean blood-pressure, has adjusted itself to a normal quantity of blood. If, then, there is a hemorrhage, there will be an additional stress laid upon the vasomotor mechanism to sustain a mean blood-pressure with a diminished volume of blood upon which to act. The mechanism, then, is at the disadvantage of having called out a part of its potential energy, leaving less with which to meet emergencies. This would be true of capillary hemorrhage, as well as of arterial or venous hemorrhage. In the latter, however, the effect is more sudden. And it was frequently noted in cases in which there was considerable loss of blood, even though the general blood-pressure had suffered but slightly and only temporarily, that such animals, so far as comparative observations permitted a deduction, did not sustain operative procedures calculated to produce shock so well as animals in which such loss of blood had not occurred.

Then, again, the anæsthetic factor is of very great importance, inasmuch as while the vasomotor mechanism is contending against the causes which tend to impair or destroy its action, if a part of the possible energy of the vasomotor system is lost by the effect of anæsthesia, it will be readily seen that to this extent the vasomotor centre is impaired, and in the cases in which the anæsthetic was too rapidly administered there was in direct proportion a rapid development of shock.

Surgical shock, then, is due mainly to a vasomotor impairment or break-down. The cardiac and the respiratory factors may be of considerable importance. However, the main effect is on the vasomotor mechanism.

If the foregoing be true, it will be seen how much more important is prevention than treatment. Prevention of shock may best be accomplished by taking into account all the known physiologic functions of every tissue and organ of the body in a way that would suggest itself to any practical surgeon. While the cause may be local, the treatment must be general. It would seem to be desirable to direct special attention to the distinction made between *collapse* and *shock.*

The result of action is reaction; of rest is restoration.

INDEX